Small Fruits
in the Home Garden

Small Fruits in the Home Garden

Robert E. Gough, PhD
E. Barclay Poling, PhD
Editors

CRC Press
Taylor & Francis Group
Boca Raton   London   New York

CRC Press is an imprint of the
Taylor & Francis Group, an informa business

Small Fruits in the Home Garden has also been published as *Journal of Small Fruit & Viticulture*, Volume 4, Numbers 1/2 and 3/4 1996.

Reprinted 2009 by CRC Press

Cover design: Donna M. Brooks.

Library of Congress Cataloging-in-Publication Data

Small fruits in the home garden/Robert E. Gough, E. Barclay Poling, editors.
     p. cm.
     "Has also been published as Journal of small fruit & viticulture, volume 4, numbers 1/2 and 3/4 1996"–T.p. verso.
     Includes bibliographical references (p.) and index.
     ISBN 1-56022-054-6 (alk. paper).–ISBN 1-56022-057-0 (pbk. : alk. paper)
     1. Berries 2. Grapes 3. Fruit-culture. I. Gough, Robert E. (Robert Edward) II. Poling, E. B. (Edward Barclay), 1953- .
SB381.S63 1996
634'.7–dc20

94-41094
CIP

Small Fruits
in the Home Garden

CONTENTS

ABOUT THE EDITORS

Robert E. Gough, PhD, is the former President of the Northeast Region of the American Society for Horticultural Science and a leading specialist in small fruit and viticulture. Dr. Gough is now Associate Professor of Horticulture at Montana State University in Bozeman. He has a diverse background, having earned a BA in English, an MS in Horticulture, and a PhD in Botany. His experiences as a county agricultural agent, state and regional extension specialist in small fruit, and a senior research scientist have provided him with a great deal of insight into the needs of growers.

Dr. Gough has published extensively–nearly 300 articles–in the area of pomology (fruit science) in both scientific journals and popular magazines, including the *Journal of the American Society for Horticultural Science, HortScience, Scientia Horticulturae, Journal of Small Fruit & Viticulture, New England Gardener, Harrowsmith/ Country Life, Country Journal, Fine Gardening, National Gardening, New England Farmer, American Fruit Grower, Proceedings of the New England Small Fruit Conference*, and *Proceedings of the Ohio Fruit Congress*. He is the author or editor of seven books on fruit culture and gardening. He also maintains an active interest in all areas relating to general horticulture, agriculture, crop science, soil science, and community gardening/landscape architecture. A member of numerous professional and honorary societies, Dr. Gough served as an Associate Editor for the *Journal of the American Society for Horticultural Science* and *HortScience* from 1985-1988. Currently, he is Editor of *Journal of Small Fruit & Viticulture*, Senior Editor for horticulture, and a consultant for Food Products Press imprint.

E. Barclay Poling, PhD, is Professor in the Department of Horticultural Science at North Carolina State University, where he has research and extension responsibilities for strawberries, grapes, and brambles. He is also the Director of the university's new Small Fruit Center. The author of numerous scientific and popular articles on strawberries and blackberries, Dr. Poling recently coauthored a book on winegrape production in the Middle Atlantic States. He has taught university courses in small fruit production and has given numerous workshops on small fruit growing to homeowner groups and master gardeners. His entertaining videos on pruning blackberries and grapes are among the most popular titles in the Agricultural Communications video library at North Carolina State University. While he is recognized internationally for his research in 'strawberry plasticulture' and plug propagation of strawberries, his secret passion is for growing blackberries! His chapter "Blackberries in the Home Garden" will surely stir considerable new interest in blackberries.

Growing Small Fruit in the Home Garden

R.E. Gough

Too often, commercial varieties of fruit have been selected for their shipping and storage qualities, and not necessarily for their flavor. This allows them to be harvested at some early stage of ripeness, shipped to your local grocery store from distant fields, and to arrive in reasonably good condition. But these fruit often don't taste very good.

There are many new varieties that have wonderful flavor but that are not suitable for commercial production. They are well suited, however, to planting in your home garden. For example, 'Herbert' blueberry is very dark-skinned but has excellent flavor. Because the market demands blueberries with light blue skin, this variety is rarely seen in commercial plantings. Some fine-flavored varieties of seedless table grapes produce plants where only 50 to 75% of the clusters are suitable for use. These are fine for the home garden, but not for the commercial grower. Growing your own small fruit will let you harvest the tastiest varieties at their peak flavor.

Small fruit take up little space when compared to tree fruit, and are well adapted to planting in the landscape. Strawberry plants make great perennial borders along walks. Grapes and the brambles form effective hedges and screens to hide unsightly areas of the

R.E. Gough is Associate Professor and Extension Horticulture Specialist, Department of Plant, Soil, and Environmental Sciences, Montana State University and Editor, *Journal of Small Fruit & Viticulture*.

[Haworth co-indexing entry note]: "Growing Small Fruit in the Home Garden." Gough, R.E. Co-published simultaneously in *Journal of Small Fruit & Viticulture* (Food Products Press, an imprint of The Haworth Press, Inc.) Vol. 4, No. 1/2, 1996, pp. 1-31; and: *Small Fruits in the Home Garden* (ed: Robert E. Gough, and E. Barclay Poling) Food Products Press, an imprint of The Haworth Press, Inc., 1996, pp. 1-31. Single or multiple copies of this article are available from The Haworth Document Delivery Service [1-800-342-9678, 9:00 a.m. - 5:00 p.m. (EST). E-mail address: getinfo@haworth.com].

landscape, or to divide the lawn area from the vegetable garden. They also make fine property markers. Currants, gooseberries, and blueberries are often used as "edible" hedges and are particularly attractive when laden with colorful fruit in summer. Sitting beneath an arbor laden with juicy grapes is a pleasant way to pass an early autumn day, and a strawberry barrel, laden and splotched red with fruit, makes an attractive conversation piece.

WHAT FRUIT TO PLANT, AND HOW MANY?

Don't waste your time and money on fruit that no one in your family wants. Gooseberries are delicious, but if no one likes them, why bother with them? Plant only what you and your family enjoy, and only enough to satisfy your needs (Table 1). Remember that just a small planting can produce an abundance of fruit to eat fresh, to

TABLE 1. Approximate yields from a planting of small fruit in the home garden.

Fruit	Ave. Annual Yield
Strawberry	70 quarts per 100 feet of row
Grape	
Vinifera	25 pounds per vine
labrusca	15 pounds per vine
Muscadine	15 pounds per vine
Gooseberry	8 quarts per bush
Currant	5 quarts per bush
Raspberry	
Black	50 pints per 100 feet of row
Red	65 pints per 100 feet of row
Dewberry	60 pints per 100 feet of row
Blackberry	
Erect	30 quarts per 100 feet of row
Evergreen	10 quarts per plant
Blueberry	5 pints per bush

freeze, and to use in making juice, pies, and preserves. Also, be careful to check the ripening schedules of the small fruit(s) you are considering against your family's summer vacation schedule!

CLIMATE

Decide which fruit you would like to have, then determine which will grow in your area. Get suggestions from your county/regional extension agent. Notice what commercial growers in your area are planting, but understand that their selection of a crop is governed not only by what will grow, but also by what will sell. Currants may grow perfectly well there, but there may be no market for the fruit and hence, no commercial plantings.

If you live in a marginal area, select only those varieties of small fruit that are adapted to your climatic conditions. For example, some new varieties of highbush blueberry grow to heights of only about 20 inches. This allows them to be insulated by snow cover in winter and so to escape severely low temperatures. Currants and gooseberries tolerate cold temperatures but not summer heat, and don't do well in southern gardens. Some varieties of grapes are extremely cold tolerant and are suitable for planting in Minnesota gardens, while "vinifera" types are better adapted to the gardens of southern California. Muscadine grapes, the vines of which bear fruit as single berries rather than in bunches, are more suited for gardens in the mid to deep south where bunch grapes grow poorly.

Heat

Most of the common small fruit–blueberries, blackberries, raspberries, currants, gooseberries, grapes, and strawberries– are widely adapted and will grow in most home gardens where summers are moderately warm, with midday temperatures of 75°F to 85°F. There are some areas, however, in which they will not perform well. Locations where summer temperatures approach 100°F for several days each year are not well-suited for most small fruit. However, there are certain varieties of *vinifera* grape that do well at that temperature. And Muscadine, or "scuppernong" grapes are tolerant of both warm and humid climates. With most of the small fruit, continual leaf temperatures above about 86°F (leaves in full sunlight can be up to 27°F warmer than the surrounding air) can cause internal water deficits, dehydration, sunburn, reduced growth, bleaching of the leaves, and a loss of fruit flavor and color. This adds to the stress of the plant and may make it more susceptible to pests. If you have many days with temperatures above this, consider types of fruit other than the common ones mentioned above, such as kiwifruit. If you expect only a few days of very high temperatures, consider sprinkling your plants. This is described later in this chapter.

Cold

Areas where winter temperatures plummet below −25°F are marginally adapted to small fruit culture. In fact, be wary of growing any small fruit if the winter temperature regularly drops below −10°F.

Mulches and heavy snow cover are beneficial and will insulate your smaller plants against extreme cold. Every inch of fluffy snow will increase the temperature around the plants by about 2°F above that of the ambient air. Small plants, such as strawberry and some of the newer half-high blueberries, can survive northern planting sites because of the insulation of heavy snow.

Many nursery catalogs list the zones in which particular varieties do well. Figure 1 is a hardiness zone map developed by the United States Department of Agriculture. This indicates the minimum winter temperatures an area of the country is likely to experience. Determine in which hardiness zone you live, then choose only those varieties that are adapted to your location. The most recent zone maps are more difficult to interpret and divide each zone into "a" and "b" areas, i.e., 7a and 7b. The "a" areas are the more northern areas of each zone, the "b" areas the more southern areas. Plants adapted to Zone 8 may not tolerate the cold winters of Zone 4, while plants adapted to Zone 4 may not tolerate the hot summers of Zone 8.

Fruit plants also must experience a certain number of hours of temperatures below 45°F in order to break their winter dormancy and start growth in the spring (Table 2). Plants in gardens too far south may not satisfy this chilling requirement and will not begin growth in the spring. If the chilling requirement is only partially fulfilled, growth will be weak and bloom spotty. Be sure to select only those plants that will be likely to satisfy their chilling requirements in your area. For example, varieties of northern highbush blueberry require a fairly long chilling period, while those of the southern highbush blueberry, or the rabbiteye, require a much shorter chilling period. Chilling requirements will be discussed more thoroughly in the individual chapters.

Fluctuating Winter Temperatures

Fully hardened, dormant plants can tolerate extreme cold better than those that are not dormant. During northern "midwinter thaws," which often occur during the second and third week of January, temperatures can sometimes approach 60°F. Because this often occurs after plants have completed their rest, the buds can respond by losing their hardiness. Blueberry tissue will begin to deharden when the temperature rises above 27°F. As temperatures rise to freezing, grape buds will deharden at the rate of about 10°F per hour, but will reharden at the rate of only about 10°F per day. This means that during the warm spell, plant tissue rapidly looses its ability to withstand cold, but only slowly regains it. When temperatures plummet after a "thaw," as they usually do, dehardened tissue is severely damaged. If you live in an area that experiences such thaws, mulch small plants heavily to minimize temperature fluctuations. There is little you can do for larger plants.

Wind

Strong winds interfere with bee flight during pollination and thus can reduce the crop, sometimes to nothing. They also stunt the plants and can dry the anthers and stigma, further interfering with fruit set and reducing the crop. Protect your planting from the strong prevailing wind by planting in the lee of a hedge, fence, or building.

Elevation

Planting on a lee hillside protects the planting from strong winds while at the same time allowing for good air drainage. Cold air,

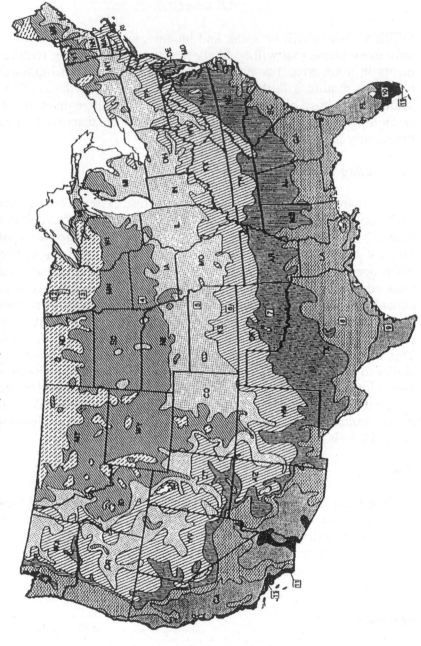

FIGURE 1. USDA map of the plant hardiness zones of the United States.

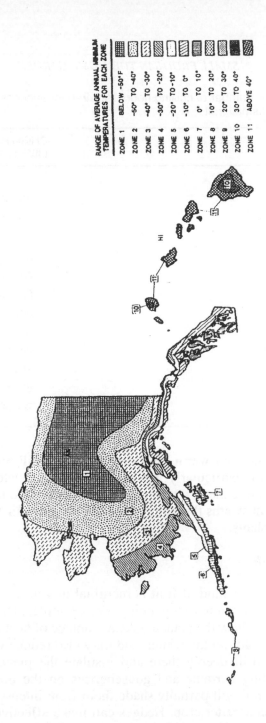

RANGE OF AVERAGE ANNUAL MINIMUM
TEMPERATURES FOR EACH ZONE

ZONE 1   BELOW -50° F
ZONE 2   -50° TO -40°
ZONE 3   -40° TO -30°
ZONE 4   -30° TO -20°
ZONE 5   -20° TO -10°
ZONE 6   -10° TO 0°
ZONE 7   0° TO 10°
ZONE 8   10° TO 20°
ZONE 9   20° TO 30°
ZONE 10   30° TO 40°
ZONE 11   ABOVE 40°

TABLE 2. The number of chilling hours required for successful growth and production of small fruit.

Crop	Chilling Hours (30°F to 45°F)
Blackberry	
Thorny	200-600
Thornless	700-1100
Blueberry	
Highbush	200-850
Rabbiteye	200-500
Currant	800-1500
Gooseberry	800-1500
Grapes	
labrusca	1200-1500
vinifera	100-400
Raspberry	800-1700
Strawberry	200-300

being heavier than warm, will roll down the hill away from the plants. This is particularly important in avoiding late spring frost damage. If a treeline or a hedge runs across the bottom of the slope, be sure to cut swaths through it to permit the cold air to drain away from your plants.

Microclimate

If a variety or kind of fruit is marginal in your area, choose the best microclimate for your planting. For example, planting next to a building will allow the plant to take advantage of heat radiated from that building during the winter, and may also reduce wind damage. Snow may drift deeply there and insulate the plant. In southern areas, planting currants and gooseberries on the north side of a building or hill will partially shade them from intense summer sun and result in a better crop. Hedges can make effective windbreaks and air temperatures on their leeward side will often be warmer than those to windward. Unfortunately, hedges, by stagnating air, can also cause frost pockets to form and actually increase the frost injury to blossoms in spring. Whenever you use a windbreak, be sure that there are openings through which some wind can pass. Good windbreaks allow about 50% of the wind to pass through. This will keep the air moving and reduce the chances of frost pockets forming.

Always plant small fruit in full sun. It is true that some kinds, like currants, gooseberries, and blueberries, can tolerate light shade, but the crop will be bigger and better under full sun conditions.

South and southwest exposures are warmest and provide the most sun. These will usually give the earliest crops, but the plants starting so early in the spring are highly susceptible to late frosts. The soil on these exposures also dries the fastest in summer, and good care in supplemental irrigation will be necessary.

Southeast exposures are slightly cooler than south or southwest exposures, since they are not exposed to direct afternoon sun. Most small fruit will do well on south, southwest, or southeast exposures.

Eastern exposures receive morning sun and are cooler than the southern exposures. The soil generally does not warm or dry as easily, and crops will be slightly later than those of plants on the more southerly exposures. Because of this, plants start later in the spring and there is less chance of late frost damage.

Northern exposures are cold and partly or completely shaded most of the day. Soil does not dry or warm readily and crops will ripen very late. They are not suited for growing any common garden crop, except possibly the cool-loving currants and gooseberries in gardens at the southern end of their range where summer temperatures would otherwise be too hot. Red raspberry aficionados living in the south may also wish to experiment with using cooler northern exposures when attempting to grow these red jewels in warmer climates.

Northeast and northwest exposures are warmer than northern but cooler than eastern or western. They are often subject to the prevailing winds or winds associated with storms and are not good choices for locating your small fruit.

Western exposures are warmer than eastern but cooler than southern. Plants here are subject to warm afternoon sun but shaded from the morning sun. Soils dry more rapidly on this exposure than on eastern exposures, and plants are often subjected to strong winds.

Wherever you locate your planting, keep it away from large trees which can shade the plants and whose roots will compete with those of the small fruit plant for water and nutrients.

SOIL

The type of soil your garden is to be planted on is one of the most important considerations in any small fruit planting. All small fruit will grow satisfactorily on well drained, loamy soils high in organic matter.

The terms "light" and "heavy," when used to describe soil, indicate the relative amounts of sand and clay that soil contains. Adding more sand makes the soil lighter; more clay, heavier. Loamy soils fall about in the middle of this category (Figure 2).

Light soils feel gritty when rubbed between the thumb and forefinger. They warm faster in the spring, provide good aeration to the root system, and do not readily compact. Plants on them start growth earlier in the spring and ripen their crops somewhat earlier. However, because plants do end their dormancy sooner on these soils, they are more subject to late spring frost damage. Also, because light soils drain easily, they are more droughty in summer and require more frequent watering (Table 3). Amend excessively light soils with the addition of peat moss, compost, and other organic matter. Plants on these will need more frequent watering and fertilizing.

Heavy soils high in clay content feel slick when moistened and rubbed between the thumb and forefinger. They warm slowly in the spring, and thus delay the start of plant growth. This helps the plants to escape damage from late spring frosts. These soils retain moisture and crops grown on them need less frequent watering. However, they can become hard and compacted when allowed to dry or are subjected to repeated heavy traffic. Amend heavy soils by adding sand, peatmoss, and compost. If water does not drain readily, install drainage tiles, or plant on small ridges to allow the roots to develop above the saturated soil.

"Shallow" soils are those that are only a couple of feet deep and

FIGURE 2. The "Soil Triangle" showing the makeup of various soil types depending upon the percentages of sand, silt, and clay they contain. Soils high in clay are too "heavy" for good plant growth and are often poorly drained and compacted. Those high in sand are too "light" and dry too fast.

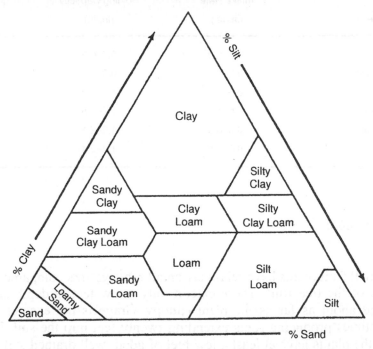

are underlain with hardpan, rock, or ledge. These are not ideal for your planting, but they can be used if amended with peatmoss, compost, and manure, and if you pay attention to proper liming and fertilizing.

If you do not have ready access to a good gardening area and soil, consider the possibility of container-fruit gardening. You can use homemade or commercial potting or bedding soil mixes to fill your containers. Bear in mind, though, that in the late fall or early winter, potted fruit plants need to be moved to the cellar or cold frame where the plants will not be exposed to potentially lethal freezing temperatures and strong winter winds.

TABLE 3. The infiltration rate and water holding capacity of various soil textures.

Texture	Intake Rate (in./hr.)	Holding Capacity (in./ft.)
Sand	1.0–5.0	0.5–0.7
Sandy loam	0.7–1.0	0.8–1.4
Loam	0.6–1.0	1.0–1.8
Silt loam	0.5–1.0	1.2–1.8
Clay loam	0.3–0.8	1.3–2.1
Clay	0.1–0.5	1.4–2.4

Water Table

Small fruit plants have relatively small root systems compared to those of the tree fruit. These extend only a few feet in depth and spread about as far as the plant canopy. Grape is the exception, sometimes producing roots extending twenty feet into the soil. Be sure the plants have at least a few feet of good, well drained soil in which to forage for nutrients. Inspect the soil of the future planting a few feet below the root zone of the plants. Well-drained soils will be a uniform, bright color–red, yellow, or brown. Mottling or gray colors indicate repeated short-term flooding. A uniform, light-gray color indicates prolonged periods of flooding. If your soil has a uniform, bright color, make a second check for adequate drainage. Dig a hole two to three feet deep and fill it with water after supper. If the water remains in the hole after breakfast the next morning, drainage is inadequate. Overly wet soils result in oxygen-starved roots and plants which are more susceptible to soil pathogens. If the soil inspection and the drainage test indicate poor drainage, choose another site for the planting. If none is available, plant on small mounds or ridges, install drains, and choose small fruit that are relatively tolerant to poor drainage (Table 4).

TABLE 4. Tolerance of small fruit to flooding and poor soil drainage.

Most Tolerant	Least Tolerant
Blueberry	Blackberry
Gooseberry	Currant
Grape	Raspberry
	Strawberry

Note: No fruit plant will do its best under conditions of prolonged flooding and poor soil drainage.

Previous Crop

Some small fruit are subject to pests they share with other plants, and should not be planted on land upon which these plants have grown within the previous five years. For example, raspberries and strawberries share susceptibility to *Verticillium* wilt with tomato, potato, eggplant, petunia, pepper, lambsquarter, pigweed, ground cherry, nightshade, horsenettle, and cocklebur. The white grub, which eats the roots of various grass species, also destroys strawberry roots. Do not plant strawberries on land previously planted to lawn or sod for at least a few years. And nematodes are particularly troublesome on sandy soil previously planted to tobacco.

Preplant Operations

Once you have selected the site for the planting, start preparing the soil. During the summer before spring planting, rototill, plow, or spade the soil, turning under any organic material that might be there.

Soil Testing

If you are planting a substantial area, take cups of soil from several representative locations around your plot, mix them thoroughly, and select a single cup of the composite sample for testing.

You may elect to have your Cooperative Extension Service or a professional soil testing service make the necessary tests, or you may do it yourself with an inexpensive soil test kit which includes analysis and fertilizer/lime recommendations for your particular soil.

Adjusting the pH

Soil pH is a measure of its acidity or alkalinity. In general, small fruit plants, like vegetable plants, do best on soil that is slightly acid (Table 5). If your soil is too acid for your crop (has a low pH), you will have to add ground limestone to raise it to a more desirable level. Sometimes, as in the case of blueberries, the soil may not be acid enough. Adding sulfur or certain other compounds will lower the pH to the range most suitable for the crop. Consult Tables 6 and 7 to determine how to correct your soil pH.

Much garden literature recommends the use of aluminum sulfate for acidifying the soil, and the compound is readily available at

TABLE 5. The range of soil pH most beneficial for the growth of small fruit.

Crop	Best Range
Blackberry	4.5–7.5
Blueberry	
Highbush	4.2–5.2
Rabbiteye	4.2–5.5
Currant	6.0–7.0
Gooseberry	6.0–7.0
Grape	
labrusca	4.0–7.0
hybrid	5.0–7.0
Muscadine	5.5–7.0
vinifera	5.5–7.0
Raspberry	5.5–7.0
Strawberry	5.0–6.5

TABLE 6. Preplant recommendations to lower soil pH to 4.5.

Change pH from	Pounds of sulfur per 1000 square feet Soil Type	
	Sand	Loam
7.5	24	70
7.0	19	60
6.5	15	48
6.0	13	38
5.5	9	25
5.0	4	13

Note: If ferrous sulfate is substituted for sulfur, multiply the amount of material applied by 6.

Adapted from: Gough, R.E. 1993. *The Highbush Blueberry and Its Management*. Binghamton, NY. The Haworth Press, Inc.

TABLE 7. Preplant recommendations to raise soil pH to 6.5.

Change pH from	Pounds lime per 1000 square feet Soil Type		
	Sand	Loam	Clay
4.0	65	175	250
4.5	55	145	210
5.0	45	115	165
5.5	30	85	115
6.0	15	45	60

most garden centers. While it will lower the soil pH, it also adds aluminum to the soil. This can build to levels toxic to the plant. DO NOT use this compound to acidify the soil. The soil pH effects plant growth in part by making some useful nutrients and minerals more available and some toxic mineral less available to the plant (Figure 3). Be sure the soil pH is correct for your crop.

FIGURE 3. The availability of various plant nutrients in the soil is governed in part on soil pH. Soils with very low pH, in the range of 4.5, often contain high amounts of available aluminum, iron, manganese, and zinc, while those with a very high pH, in the range of 8.0, contain high amounts of calcium and molybdenum. Most small fruit plants do best in a slightly acid soil with a pH of 6.0 to 6.5 where there is a more moderate nutrient availability.

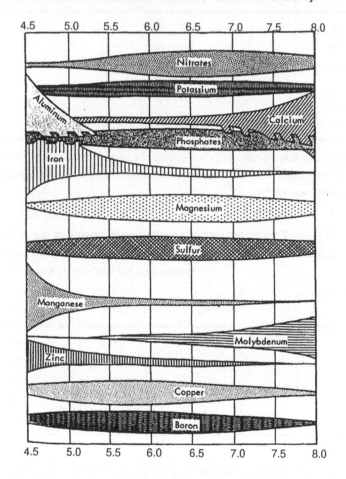

Green Manure

After correcting the soil pH, plant a summer green manure crop, such as buckwheat or millet. Such crops are temporary. When growing, they crowd out weeds and, when plowed under, add valuable organic matter to the soil. Add enough complete fertilizer, such as 10-10-10, or manure, to produce good cover crop growth. This will also reduce nitrogen depletion when the crop is turned under.

Plow down the summer crop in early fall and plant a winter crop, such as winter rye. Also, add manure, compost, and other organic material to the plot at this time. This is one of the last times that you will be able to strongly influence the organic matter content of the soil.

In spring, turn under the winter crop at least two weeks prior to planting to allow the soil to settle somewhat.

Small Areas

If you will plant only a couple of grapevines or berry bushes, you needn't prepare a large area. Prior to planting, dig out large planting holes at least three feet wide and three feet deep and refill them with a 50/50 mixture of compost or peatmoss, and soil. Adjust the pH of this mix to fall within the range suitable for your plants. You should adjust the pH of your soil two seasons before planting, that is, in the fall prior to spring planting, and in the spring prior to fall planting.

PLANTING

Time

Early spring is the usual time for planting most small fruit crops in most areas. This will allow the plant enough time to establish a good root system before winter, which will in turn reduce the chances of winter damage from desiccation or plant heaving.

In the south, where winters are not severe, small fruit can be planted in the fall, which will allow the plants to establish good root systems before the onset of the hot summer. Raspberries do especially well when suckers are planted in the fall in the south. Because they bloom so early in the spring, currants and gooseberries are often planted in the fall, even in the north, to allow the plant time to establish a root system before it begins its growing season.

Digging the Holes

Be sure to dig the planting holes deep enough and wide enough to accommodate the entire root system of the plant without bending or twisting. If some roots are excessively long, cut them to fit. Twisting them to fit the hole will stunt the plant and cause it to be shallow rooted. Wrapping them around the inside of the hole could also cause them to girdle and kill the plant over time.

If you've dug the holes with an augur or posthole digger, the sides may become compacted. This is particularly true on heavy soil. Scrape and loosen the soil on the sides of the hole with a spade before planting.

Purchasing Plants

Purchase only healthy, virus free small fruit plants from a reliable nursery. Usually it's better to purchase from a nursery near your home to avoid delays in transplanting, although local nurseries may not have the variety of plant that you want in stock. If ordering from a mail-order nursery, be sure to place orders at least six months in advance of planting.

Some small fruit plants are now being propagated by tissue culture (TC). This method allows better control of virus infection. Virus infected plants are less vigorous, have a shorter productive life, and may produce smaller fruit. So, spend a little extra money to get the best virus free plants available.

Good small fruit planting stock should be vigorous and have a heavy, dense root system. Most kinds are sold as one year old plants, though blueberries should be purchased as two year old (eighteen to twenty four inch high) plants.

Inspect your plants upon arrival. If the packing material is dry, moisten it. If the roots are moldy or the plants partially decayed, notify the nursery immediately. Store the plants in a cool, damp area until planting.

Pollination

Most small fruit are self-fruitful, that is, they will set a crop by their own pollen. However, you'll get more, bigger, better, earlier ripening fruit if you plant at least two different varieties of each fruit.

Most small fruit plants are pollinated by bees. You may have to rent hives if your natural populations are low. Also, be sure to keep dandelions and other wild flowers from blooming when your small fruit are in bloom, since bees often prefer the wild flowers to those of your fruit.

Specific pollination requirements are considered in the various fruit chapters.

Removing the Flower Buds

Removing the flower buds at planting time on small fruit bushes will prevent them from bearing a crop the first season. Bearing fruit so early in the life of a plant could stunt its development and reduce its capacity for bearing substantial crops later on. Remove flowers on strawberry plants at bloom.

Planting

Spread the root system in the hole at roughly the same depth at which the plant grew in the nursery, and begin filling with good topsoil or a 50/50 mixture of compost (or peatmoss) and topsoil. Never put plain peatmoss into the hole, particularly if it is dry, since it could soak up moisture from the plant roots.

Fill the hole half way with the mix, tamp firmly, and flood with water. As the water drains, it will settle the soil around the roots. Fill the remainder of the hole, tamp firmly, and flood again to settle the soil.

The purpose of tamping and flooding is to eliminate air pockets from the root zone. These can dry the roots and may harbor diseases. These steps also bring the roots into intimate contact with the soil to allow for easier extraction of water and nutrients.

Do not place any lime, bone meal, or granular fertilizer into the planting hole. Lime and fertilizer could burn the roots, and bone meal is of questionable value. If you wish to give the plant a "boost," water it in with a commercial starter solution or a manure tea, but if your soil was prepared properly this is not necessary.

MANAGEMENT

Cultivation

Weeds compete with small fruit plants for water, nutrients, and sunlight. They can also harbor insects and diseases that could infest the crop plants. Cultivate frequently enough to control weed growth. Simple hand hoeing, with a sharpened hoe, is very effective. Hand hoeing is done properly with a scraping and not a chopping motion. Chopping the soil will leave it rough and more difficult to hoe the next time, and could damage the root systems.

Many gardeners run a rototiller down the rows every week or so to control weeds and to give the garden a "dressed" appearance. This constant tilling chops roots, destroys the soil structure, and increases soil aeration, which in turn promotes drying, loss of organic matter, and compaction. Hoeing is better for the plants and the soil, but it does have some drawbacks.

Hoeing is repetitive and time consuming. It leaves a bare soil that often is either dusty or muddy, and will more easily compact as you move through the planting. Fruit will in turn be dust-covered or mud-splashed and therefore less attractive. Temperatures of bare soil fluctuate widely, and the soil dries more rapidly.

Mulching

Mulching is done once a year or once every two years, usually in the early spring when fruit are not present. Mulch is more pleasant to walk on than is bare soil, and will make the fruit neither dusty nor muddy. Mulch holds soil moisture and organic mulches improve the soil structure. Soil beneath a mulch will not compact easily, remains fluffy, and maintains good moisture and aeration. Temperatures remain up to 20°F cooler in summer and warmer in winter under a mulch and do not fluctuate widely. This promotes better root growth and reduces the chances of frost heaving. In addition, organic mulches add valuable organic matter to the soil as they decompose. Yields and size of most small fruit are increased substantially in mulched plots. In the home garden, mulching is wiser than cultivating or hoeing.

Black polyethylene can be used effectively as a mulch, particularly with some strawberry systems. This material is inexpensive, easy to apply, and warms the soil to promote faster root growth. However, it may also harbor mice and snakes, and does little to improve soil structure. Further, black plastic can result in very shallow rooting which makes the plant more susceptible to drought.

Because it breaks down into small pieces after a couple of years, it may also be difficult and messy to use around long-lived perennials.

Organic mulches are somewhat more difficult to apply but add organic matter to the soil upon decomposition.

Wet, shredded newspaper makes an effective mulch for many plants, except strawberries, but usually breaks down within a single growing season. It is flammable and can blow away when dry.

Straw is also an effective mulch. It is usually inexpensive, does not mat down and does not blow away once it is wet. Like paper, it is flammable and might also harbor mice. Mice can damage your plants by eating the bark during winter.

Salt hay is cut from marsh grass on coastal salt marshes. It makes a good mulch because it is generally free from seeds that can germinate under garden conditions. Like straw, it is flammable and could harbor mice.

Hay is not a good mulch. It usually contains too many weed seeds, is flammable, and will harbor mice.

Cornstalks make a good, coarse mulch, as do ground corn cobs.

Leaves from yard and street trees are effective but can mat down and interfere with infiltration of water and air, as well as blow away. Mix evergreen bows with the leaves to keep them from packing and to hold them in place. Do not use maple or elm leaves around blueberry bushes, since they leave an alkaline residue upon decomposition.

Buckwheat and peanut hulls make an effective and attractive mulch, but their cost might be prohibitive.

Pine needles make an excellent mulch. They do not blow away or compact, and they rot very slowly. They are reasonably priced at several dollars per bale, or you can rake them yourself.

Wood chips and sawdust are also excellent mulches. They rot slowly, are usually available free in large quantities, and do not harbor mice to a great extent. Once wet, they will not blow away. Be sure to obtain your sawdust from a sawmill and not from a carpenter's shop. The fine dust from the latter will crust and interfere with air and water infiltration. Many garden books claim that sawdust will acidify the soil. This is not true. Soil pH remains nearly constant under rotting sawdust mulch.

Few studies have been done on the efficacy of hardwood vs.

softwood sawdust. Softwood sawdust decomposes more rapidly than hardwood and contains more pitch and resin. Maple sawdust may raise the soil pH upon decomposition, and black walnut sawdust may contain compounds toxic to plants. Practically speaking, however, most sawdust is from "mixed" hardwood/softwood species and presents little danger to small fruit plants.

Apply organic mulches immediately after planting and to a depth of at least four inches. Six inches is better. If plants are shorter than this, as in the case of strawberries, slope the mulch into the plant. Spread the mulch completely around the plant and keep the area beneath the dripline mulched as the plant grows. Keep the surface of the mulch level or sloped slightly toward the plant to prevent water from running away from the root system.

Microorganisms use some soil nitrogen to decompose organic mulches, competing with the plant for this nutrient. Whenever you use an organic mulch, add additional nitrogen fertilizer to compensate for that used in decomposition. The amount varies according to the mulch—the woodier the material, the more nitrogen you will need. Woodiness can be expressed by the carbon to nitrogen (C:N) ratio (Table 8). The higher the C:N ratio, the more woody the material. In general, add about two pounds of 10-10-10 fertilizer to each 100 pounds of mulch at the time of mulch application. Check plants often for yellowing of the older leaves and stunted growth, both signs of nitrogen deficiency.

PRUNING

Pruning is the removal of unproductive, diseased, or dead wood. By doing this on a regular basis, you reduce the number of pest problems as well as make room for more productive new wood. Allowing better light and air penetration of the bush will increase its productivity and decrease disease incidence.

Pruning also allows you to shape the plant and to train it to a desirable form. For example, grapevines must be pruned heavily each year and trained to a trellis or arbor to produce the best fruit. If this is not done, they will degenerate quickly into a tangled and unproductive mass of wood. Drooping blueberry canes should be pruned out to keep the bush from falling into the aisles.

TABLE 8. Approximate Carbon:Nitrogen (C:N) ratios of some common materials.

Material	C:N Ratio
Wood chips	600:1
Sawdust	300:1
Paper	125:1
Timothy	80:1
Corn Stalks	80:1
Oat Straw	75:1
Autumn Leaves	65:1
Horse Manure	35:1
Rotted Manure	20:1
Cow Manure	18:1
Table Scraps	15:1
Grass Clippings	15:1
Normal Soil	10:1
Hen Manure	7:1

Pruning can remove excess flower buds and thus reduce competition among developing fruit. This increases their size.

The process of pruning is both a dwarfing and a stimulating operation. By removing unproductive wood, you stimulate vigorous new growth. However, all the new growth usually does not equal the total amount of old growth removed. This helps to keep the plant in bounds.

Time of Pruning

Begin pruning at planting by clipping excessively long roots and by removing dead or damaged shoots, and flower buds. Small fruit plants usually require little pruning thereafter until their second or third year in the field.

Severity

Prune lightly until the plant begins to bear. Heavy pruning while the plant is still young will keep it overly vigorous and can delay the onset of bearing.

Once the plant is mature, prune only enough to remove damaged and unproductive wood, and to keep the plant in bounds.

Season of Pruning

It makes little difference to the plant when, and if, it is pruned. However, some seasons are better than others from our standpoint. Late winter is the traditional time to prune small fruit. There is no foliage to obscure the plant's framework, no flowers or fruit to knock off, and few other chores to do around the home. It is cooler and more pleasant to work then, and most winter damage has already occurred and can be pruned out. The roots have nearly their full complement of nutrients in storage, and pruning cuts will not be subject to disease and extreme cold. For all these reasons, it makes sense to prune in late winter.

Some gardeners like to prune in the late fall. This is usually satisfactory in areas that do not experience a severe winter, since very cold weather can damage the cambium tissue just beneath the bark at the edge of a pruning wound. Early fall pruning, before the plant has entered dormancy, is not advisable, since it could stimulate late growth which may not harden off in time for winter. Such succulent growth will winterkill and may become the point of entry for pathogens. Winterkilled tissue must be pruned out in late winter, thus making two prunings necessary. Early fall pruning will also deprive the plant's root system of some nutrients that would ordinarily be transported to the roots from the top of the plant. This could weaken the plant and make it more susceptible to stressful conditions.

Spring or summer pruning is unsatisfactory since the plants have leafed out and the framework is obscured by foliage. Buds, flowers, and fruit are also easily injured during these pruning operations.

Grapevines will bleed profusely if pruned at anytime other than late winter. While this does not harm the plant, it can ruin the aesthetics of a brick walk beneath the arbor.

Special Pruning Operations

Sometimes, grape growers thin the berries and the clusters to increase the size and appearance of the fruit. Do this according to directions in the grape chapter. Blueberry growers sometimes thin the number of flower buds to increase the size of the remaining fruit. This is seldom necessary under most conditions.

Wound Dressings

Wound dressings primarily soothe the psyche of the gardener. They do little to aid the healing process and can sometimes even retard it. Don't use them.

FERTILIZING

Plants and berries are made up of carbohydrates and minerals. The minerals originate in the soil, from which they are extracted by the root system and transported throughout the plant. There they remain until the tissue dies and rots, thence to be returned to the soil. If the tissue is removed from the site, as during pruning and harvest, the nutrients are deposited elsewhere. For example, a 100 foot row of blackberries yielding about 350 pounds of fruit will accumulate about fourteen ounces of nitrogen, 4.7 ounces of phosphorus, and fourteen ounces of potash in the berry tissues. With yearly removal of the crop and the prunings, about sixty five ounces of nitrogen, nineteen ounces of phosphorus, and forty two ounces of potash are taken off the land. Similarly, the crop of a single grapevine producing about fifty pounds of fruit removes about 1.15 oz of nitrogen, 0.76 oz of phosphorus, and almost two ounces of potash. When added to the nutrients in the prunings, 2.7 oz of nitrogen, 1.15 oz of phosphorus, and 3.2 oz of potash are removed from the soil each year. Over time, the soil upon which the plants are growing is drained of its minerals. These must be replenished periodically to keep the plant in good health. That is the role of fertilizers.

Small fruit plants need varying amounts of up to sixteen elements. Three of these—carbon, hydrogen, and oxygen—they get in unlimited amounts from the air and water. Of the remainder, only nitrogen, phosphorus, and potassium are used in large amounts and are most often deficient. For this reason, most common commercial fertilizer contains a blend of only these three elements. That blend is expressed as three numbers on the package. This is called the fertilizer "analysis." A package labeled "10-10-10" contains 10% nitrogen, 10% phosphorus, and 10% potassium (potash). The elements are always given in the same order, as N-P-K. So, if a fifty pound bag of 10-10-10 fertilizer contains 30% nutrients (10% + 10% + 10%), what is the remaining 70%? Filler! Since you cannot spread pure forms of these elements, they must be mixed and diluted to be useful. The higher the analysis, the more nutrients and the less filler you apply.

Commercial vs. Organic Fertilizers

Organic fertilizers are produced naturally from plants and animals. Manure is a good example. These tend to be bulky and difficult to handle. Many of them have an odor that is offensive to some, and all have a relatively low analysis (Table 9).

The use of commercial fertilizers exploded after World War 2. These are made of minerals mined and ground and concentrated, or from byproducts of the oil industry. They are relatively inexpensive, have no odor, and are easy to handle. Because their analyses are much higher than those of organic fertilizers, you have to use much less of the product to get growth comparable to that from application of organic fertilizers.

The subject of commercial vs. organic fertilizer is fraught with emotion and misinformation. To label all commercial fertilizers as "bad" and all organic fertilizers as "good" is silly. There are "good" and "bad" components in each category. For example, as we mentioned above, some organic fertilizers are bulky, difficult to handle, smelly, and low and sometimes uncertain in analysis. But because they have a low analysis they are less apt to "burn" the plants. Their bulky nature is due to the presence of large, organic molecules. These are useful in "fluffing" the soil and in increasing its aeration and drainage. Because the organic matter soaks up water, it reduces leaching. This is the reason behind adding peatmoss or compost to the soil. The plant cannot take up the nutrients in these large molecules. Slowly, microorganisms in the soil and in

TABLE 9. Proximate composition of some animal manures.

Animal	Moisture (%)	Proximate Composition*		
		N	P	K
Fresh Manure/Bedding				
Cow	86	11	4	10
Duck	61	22	29	10
Goose	67	22	11	10
Hen	73	22	22	10
Hog	87	11	6	9
Horse	80	13	5	13
Sheep	70	20	15	21
Turkey	74	26	14	10
Dried Commercial				
Cow	15	30	40	25
Hen	13	31	35	40
Hog	10	45	42	20
Rabbit	16	26	31	32
Sheep	10	32	25	41

*pounds per ton

the manure itself break down the molecules to release nutrients in the simple form the plant can absorb. Because this process can take several months, organic fertilizers are sometimes called "slow-release." This also makes them less apt to burn the plant. Their slow release nature is useful in long term feeding but is not useful when the plant needs a quick "boost" to eliminate some nutrient deficiency. Improper application of organic fertilizer can also stimulate excessive growth late in the season which will not harden in time and subsequently winter kill.

Commercial fertilizers are compact and concentrated. A fifty pound bag of 10-10-10 contains the same amount of nitrogen as 500 pounds of cow manure. Commercial fertilizers are essentially odor-less, spread easily by hand or in a spreader, and have a guaranteed analysis. Manure and some other organic fertilizers, because of their age or the way they were handled, have an uncertain analysis. The nutrients in commercial fertilizers are contained in small mole-cules that the plant can readily absorb. This makes them fast-acting. While their application can stimulate rapid growth, it is this same attribute that can make them "burn" the plant if they are applied in clumps or in excessive amounts. The "burning" is actually caused by the fertilizer salts removing water from the tissues. A similar thing happens to us when our lips shrivel after we have eaten too many salted nuts. Because their filler is inorganic material with little nutrient value, commercial fertilizers add no organic matter to the soil. They cannot bulk up the soil, nor increase its water and nutrient holding capacity. In fact, their constant and exclusive use leads to more rapid breakdown of the soil organic matter and a decrease in soil tilth.

The use of organic fertilizers to the exclusion of commercial fertilizers has become a gardening philosophy to some. As such, it is fruitless to argue its problems. If you wish to garden organically, go ahead, but understand its shortcomings, its dangers, and the amount of hard work it involves. Understand too that, as far as the plant is concerned, there is absolutely no difference between a nitrate ion absorbed from 10-10-10 and one absorbed from rotten blood. In my opinion, the wise gardener will use a combination of organic and commercial fertilizers since there is no apparent advan-tage to using the more expensive organic fertilizers.

Amounts

Regardless of whether you use organic or commercial fertilizer, you must apply the correct amount for best growth. Apply too much and you will stimulate rampant vegetative growth. This will be soft, more prone to pest attack, and may not harden sufficiently before winter. Fruit from these plants will also be soft, poorly-colored, bland-tasting, prone to rots, and highly perishable. Apply too little fertilizer and plant growth will be stunted, the leaves small and yellow, root growth poor, and the fruit small and off-colored.

Application

Apply fertilizers in a broad band beneath the dripline of the bush and completely around it. Fertilizing on only one side of the plant will encourage growth only on that side. Strawberry is an exception.

Apply fertilizers as suggested in the chapters, but apply manure only in the fall after the plants have entered dormancy but before the ground has frozen.

LIMING

Keep the soil pH adjusted to fall within the proper range for your fruit. Since lime is relatively immobile in the soil, incorporate it well into at least the top foot of soil before planting. The same is true for phosphorus.

Use ground agricultural limestone, commonly called lime, or dolomitic lime, which, in addition to calcium, also contains magnesium, another important nutrient. Do not use hydrated lime.

Soil pH and Nutrients

Soil pH affects the availability of nutrients and must be maintained at the proper level. In most cases, this is not a problem. But a problem will develop with blueberry if the soil pH rises above about 5.5. At that range, these plants can no longer absorb iron efficiently from the soil. The result is a yellowing of the newer leaves called iron chlorosis. The lemon-yellow color can be corrected temporarily by spraying iron chelate on the plant, but lowering the soil pH is the only way to correct it permanently.

WATERING

Fruit plants need at least an inch of water per week during the growing season. If the weather is hot and the fruit load heavy, they may need several inches. Always water before the plants begin to wilt.

Apply water to the soil beneath the plants, where it can soak in and reach the roots quickly. Soaker hoses are commonly used for this. Sprinklers waste water through misplacement and evaporation, although sprinkling the plants during extremely hot weather will cool them sufficiently to promote better growth.

However you choose to water, water deeply, and wet the soil to a depth of at least six inches. Simply sprinkling the plants to wet the surface of the soil only wastes water and time.

If you use overhead sprinklers, be sure to do so only when the temperature is rising. This will give the foliage a chance to dry before nightfall and so lessen the chances of disease.

PESTS

All gardeners, organic or commercial, conscientious or lax, have pests. But you can minimize your problems by planting varieties resistant to some pests in your area, by keeping the planting clean and free of weeds, trash, and diseased wood, and by taking immediate control measures once you identify a problem.

Hand pick insects, remove dead and diseased canes at once, and rogue any plant you suspect has a virus. Spray if you must, but keep it to a minimum.

Chemical pest control can be as emotional a topic as commercial fertilizers. Unfortunately, sometimes there is no other way to control a pest except through application of a pesticide. Here too there is much information and silliness being spread around by people who should know better. First, there is no such thing as a non-toxic pesticide. A pesticide kills pests, so it must be toxic. There are, however, varying degrees of toxicity. Some formulations are relatively non-toxic, some highly toxic. Some compounds are highly toxic to some insects but relatively non-toxic to others.

Organic gardeners will use only organic pesticides, some in the belief that, because they are natural they are "safe" and will not harm the environment. Some of these are quite effective, others are useless. But to be effective, all must be toxic to something. In fact, rotenone, an organic insecticide, is relatively non-toxic to people but extremely toxic to fish. Nicotine, a natural compound used extensively by some gardeners, is highly toxic to bugs and the gardener. As little as a tablespoon or so will kill a horse. On the other hand, some commercial pesticides, like Guthion, are highly toxic to people and pests, while others, like carbaryl, are highly toxic to bees but relatively non-toxic to people. Like the issue of fertilizers, there are "good" and "bad" pesticides on both sides. Use pesticides only when you must, and then use only the most effective and least toxic ones available. Always follow directions on the package carefully.

HARVEST

Small fruit are all soft fruit and will not keep well for long periods of time after they are ripe. Rotted fruit left in the garden will harbor insects and diseases that can spread to other fruit.

Harvest ripe fruit on a regular basis. If there are too many to use immediately, consider preserving them by freezing or processing into jellies, jams, or wines.

Instructions for harvesting the specific fruit are given in the following chapters.

The small fruit planting is an exciting area of the home landscape. Nearly everyone has a vegetable garden with the same old tomato and cucumber plants that everyone else has in their vegetable garden. But the small fruit grower has something different; something nutritious and beautiful to look at that does not take up all your spare time. It is something that not all your neighbors have, and you'll personally reap hours of enjoyment in its care and in its harvest. I hope this volume will excite your imagination and encourage you to plant small fruit in your home garden.

Blackberries

E. Barclay Poling

Many people still think of blackberries as growing only in the wild. Fortunately, plant breeders affiliated with the U.S. Department of Agriculture and several Land Grant Universities have successfully introduced "tame" blackberries for the garden or home landscape. Even the objectionable thorns or prickles have been "bred out" of some of the newer erect blackberry varieties, including 'Navaho' and 'Arapaho.' The introduction of these newer blackberries for domestic culture is well-timed, as many of the woods and field fence rows desirable for 'blackberrying' have given way to urban development. The "tame" blackberry varieties for the garden may not compare perfectly in flavor to the wild berries of your memory, but at least you will not have to wander any farther than your backyard to collect them, and without the company of ticks and chiggers!

Growing your own patch of blackberries is relatively easy, and it may be the only way to assure a home supply of these delicious fruits. Unlike raspberries, there is only a limited amount of commercial blackberry production in the U.S. When you see fresh blackberries in the supermarket or grocery store, the price is often

E. Barclay Poling is Professor and Extension Specialist of Small Fruits, Department of Horticultural Science, North Carolina State University.

The section on Trailing Blackberries (pp. 57-60) is adapted from: "Growing Erect and Trailing Blackberries" by the late G.M. Darrow (1948), U.S.D.A. Farmers' Bulletin No. 1995. pp 14-20.

[Haworth co-indexing entry note]: "Blackberries." Poling, Barclay E. Co-published simultaneously in *Journal of Small Fruit & Viticulture* (Food Products Press, an imprint of The Haworth Press, Inc.) Vol. 4, No. 1/2, 1996, pp. 33-69; and: *Small Fruits in the Home Garden* (ed: Robert E. Gough, and E. Barclay Poling) Food Products Press, an imprint of The Haworth Press, Inc., 1996, pp. 33-69. Single or multiple copies of this article are available from The Haworth Document Delivery Service [1-800-342-9678, 9:00 a.m. - 5:00 p.m. (EST). E-mail address: getinfo@haworth.com].

33

sky high! You may even be motivated to consider the possibility of growing blackberries for *profit!* But, a word of caution about the perishability of blackberries—freshly picked, these berries have the shortest shelf-life of any small fruit. Therefore, it may be advisable to limit your initial distribution to immediate neighborhood friends and pick-your-own customers!

As I have mentioned, blackberries are not a specialty crop requiring the skills of an expert gardener. In fact, they are probably the easiest and least expensive small fruit to grow. You will need little more than a small area that is open to the sun, and a soil that is reasonably well-drained. A 10-foot hedgerow of erect blackberries can produce more than 30 pounds of fruit! You can establish it with as few as 4 blackberry plants (spaced 2.5 feet apart), and your cost for plants will be less than 2 *Blockbuster*™ movie rentals! The lapse time between establishing a blackberry planting and having some berries to enjoy can be as little as one full year; and you can expect full production in the second year following planting. With good horticultural and pest management practices, the life expectancy of an erect blackberry planting can be 10 to 12 years; I am aware of healthy semi-trailing thornless blackberry plantings that are nearly 20 years old.

Blackberries can be used in a variety of creative ways in the home landscape. The upright blackberries (erect types) are great for forming effective hedges and screens to hide unsightly telephone or electrical boxes. They can also be used for showing a property line or simply dividing your lawn area from the vegetable garden. A trellis for supporting the trailing or semi-trailing type blackberries can provide the perfect backdrop for almost any landscape setting. In the spring, the white and pink blooms of the semi-trailing thornless blackberries will have great ornamental value. While travelling in Holland one summer, I came upon a semi-trailing thornless blackberry, 'Black Satin,' being trained on an espalier against the side of a lovely brick home. Sometimes, rather than constructing a trellis, the semi-trailing thornless blackberry plant can simply be allowed to grow along a split rail fence. The only limit is set by your imagination!

If you have a little spare time, you may wish to read through the information in the next section on the origin, history, and botany of blackberries. Otherwise, if I have overstimulated your appetite for the "nuts and bolts" information on growing blackberries, you may skip ahead to the information on selecting, planting, pruning, training your blackberry plants, and controlling their pests.

CLASSIFICATION AND ORIGIN

Blackberry plants have relatively flexible woody stems that benefit from external support (stakes, trellis, or lateral support wires). They run the complete range in growth from upright (erect) types that can be grown without any external support, to the semi-trailing and trailing types that must be trellised or staked for support of the relatively heavy fruit load that mature plants are capable of producing. Distinctions may also be made in reference to the relative plant size and fruit bearing capacity of the blackberries, but we will postpone this discussion until later in the chapter.

Blackberries belong to the very "cosmopolitan" Rose family (Rosaceae), the same family that includes apples, peaches, plums, roses, strawberries and raspberries. Actually, blackberries, dewberries, and raspberries (all "brambles") belong to the same genus, *Rubus*. *Rubus* is a very diverse and widespread group of plants—it is estimated that there are between 400 and 500 individual raspberry and blackberry species in North America, South America, Hawaii, Europe, Africa and Asia. One sure way of distinguishing a raspberry from a blackberry is to carefully observe how the fruit separates from the bush. Unlike the raspberry, but like its cousin the dewberry, the blackberry fruit has a solid core or receptacle. At harvest, raspberry fruits separate from the receptacle and their cores are not eaten. Put another way, the raspberries are hollow and thimblelike when picked, whereas the surface "drupelets" of blackberry and dewberry adhere to the receptacle. In all brambles, the real fruits are actually separate miniature drupes or drupelets each with a single tiny seed—these drupelets adhere to one another to form what is called an "aggregate fruit." *Rubus* is further divided into two subgenera, *Idaeobatus* (raspberries) and *Eubatus* (blackberries), based on whether the fruit separates from or includes the core at harvest.

North America has many native blackberries and dewberries.

Originally, when North America was heavily forested before the settlers arrived, there were only a few distinct species of blackberries. But, as forests were cut and land was cleared, blackberries spread and there was an opportunity for seedlings of different species to grow side by side. Bees and other insects cross-pollinated the plants, and birds assisted in this vast natural breeding program by "dispersing" the blackberry seeds throughout the countryside. Blackberry plants were occasionally transplanted to early American gardens, but most of the agricultural books and papers of the day were chiefly focused on how to destroy the plants and keep them from starting again!

As time went on, researchers discovered wild hybrid blackberries which became the foundation for our modern garden blackberry varieties. Interest in domesticating wild blackberries in America dates to about 1850. There are several noteworthy native blackberries and dewberries from which modern varieties have been developed. *R. allegheniensis* Porter is the eastern North American erect blackberry with "stout" prickles on its canes. Many modern varieties, such as Eldorado, Darrow, and Cherokee, are derived from this species. Eastern North America is also the center of distribution for *R. baileyanus* Willd., the American dewberry or trailing blackberry—'Lucretia' is one variety derived from this species. Dewberries are basically distinguished from blackberries by their trailing growth habit; cultivated blackberries are erect plants. As marketed, blackberry and dewberry fruits cannot be easily distinguished. The name dewberry is thought to have been given because the berries were covered with dew when gathered.

Some of the most important dewberry varieties of commerce originated from the hybridization of raspberries and blackberries. One of the most famous early products of such a cross was 'Loganberry,' a vigorous grower of long-trailing habit that is not grown in the East on account of lack of hardiness, but is still grown on the Pacific Coast. Another raspberry/blackberry hybrid, 'Phenomenol,' was later crossed with a dewberry to produce 'Young' and 'Boysen.' Tayberry was produced by the hybridization of blackberry and red raspberry in Scotland.

BLACKBERRY TYPES

There are 5 different types of blackberries found in the U.S. (Table 1), but in this chapter, I will concentrate on the eastern erect or nearly erect types, the semi-trailing thornless blackberries, and the eastern trailing blackberries, or dewberries.

As noted earlier, the erect blackberries are upright and self-supporting, the semi-trailing types are free-standing at the base with the top portion arched over, and the trailing blackberries (dewberries) have weak canes that must be tied to poles or trellises. Of the three types, the trailing blackberries are usually the earliest to ripen (soon after strawberry season), and the erect blackberries can begin ripening a week or so later, depending on variety. Semi-trailing thornless blackberries usually do not begin ripening until mid-summer. Dewberries are typically short seasoned and low yielding compared to the erect and semi-trailing blackberries. However, the trailing blackberries are generally sweeter than the erect blackberries. The harvest of erect blackberry types normally extends 2 to 3 weeks; it is unusual to pick any given thornless type for more than one month. In the third year, yields of erect thorny varieties such as 'Shawnee' can be as high as 5 or 6 pounds per plant, and yields in excess of 30 pounds per plant are not uncommon for the semi-trailing thornless blackberries. However, semi-trailing blackberry plants are spaced 8 to 10 feet apart whereas erect blackberry plants or root cuttings are set about 2 feet apart in the row. As a rule, the thornless semi-trailing blackberries are less flavorful than the erect types.

HUMAN NUTRITION

Fresh blackberries are highly nourishing. They contain 85 percent water and 10 percent carbohydrates, in addition to a high content of minerals, vitamin B, vitamin A and calcium, which is present in blackberries more than in any other fruit. You can also enjoy blackberries in countless other ways: frozen, in cobblers (my favorite), in preserves, juice, jam, pie, sauce, pancakes, pudding, shortcake, ice cream toppings, and yogurt flavorings.

TABLE 1. The blackberries of North America fall into five major groups.[1]

Type	Region	Examples(s)
1. Erect or nearly erect	Eastern U.S. from Florida to Canada and from the Atlantic Coast to the prairie states.	Cheyenne, Cherokee, Shawnee, Navaho, Arapaho
2. Eastern trailing blackberries	Approximately the same range as the erect or nearly erect blackberries.	Lucretia
3. Southeastern trailing blackberries	Ranging along the Atlantic and Gulf Coasts, from Delaware to Texas.	Oklawaha
4. Trailing blackberries from which 'Logan' is derived	Ranging along the Pacific Coast from Canada to S. California	Loganberry, Nessberry
5. Semi-trailing thornless	Adapted to eastern U.S.	Black Satin, Chester

[1] Adapted from George M. Darrow, In: Blackberries-Dewberries, Fruit and Vegetable Facts and Pointers, Dec. 1958.

ADAPTATION

There are "tame" blackberry varieties for practically every section of the U.S., except the extreme northern states. But generally, the blackberry is best suited to the western and eastern seaboards and the southern two-thirds of the U.S. Blackberries, more than raspberries, are quite tolerant of the heat and drought conditions that frequently prevail during the summers in the southern states. In Florida and extreme southern Texas, the chief climatic limitation for certain "higher chill" erect thorny blackberries, including 'Cheyenne' and 'Shawnee,' is insufficient winter chilling. Certain southern trailing dewberries are limited to subtropical regions because of extremely early spring bloom and resultant frost hazard. In several publications, erect thorny blackberries and semi-trailing thornless blackberries are mentioned as being equally susceptible to cold injury, but my experience in North Carolina is that the erect thorny blackberries are the more hardy of the two. The semi-trailing thornless varieties may be injured when the temperature falls below 0°F; the 'higher chill' erect thorny blackberries (from the University of Arkansas and the University of Illinois) can withstand winter temperatures down to $-15°F$ ($-26°C$). The trailing blackberries or dewberries are comparable to the semi-trailing thornless blackberries in cold hardiness. If you live in a region that is borderline for semi-trailing thornless blackberries and dewberries, cover the canes with mulch during winter and the buds and canes will survive temperatures lower than 0°F ($-18°C$). Some of the newer semi-trailing blackberry varieties, including 'Chester Thornless,' have demonstrated improvement in cold hardiness as well as improved flavor, color stability and shelf-life.

Some of the erect blackberries, including 'Brazos,' 'Rosborough' and other Texas varieties, are also vulnerable to winter injury when grown in areas where temperatures go as low as 0° to 5°F (18° to $-15°C$) these less hardy erect blackberries will be noted in the next section on "Plant selection."

Of course, susceptibility to freeze injury is variable depending upon winter preconditioning. Certain southern trailing types are limited to subtropical regions because of extremely early spring bloom and resultant frost hazards. These include the Florida varieties 'Flordagrand' and 'Oklawaha,' which are no longer commercially propagated.

Disease pressures seriously limit the area of adaptation of many of the erect and semi-erect blackberry varieties. Rosette or double blossom, a fungal disease, and sterility, a virus complex, can severely limit production in coastal areas. Rosette is a serious problem throughout the South, and is reported to have almost eliminated commercial blackberry production in Florida.

VARIETY SELECTION

There are relatively few blackberry varieties that can withstand winter temperatures below −10°F (−23°C). Breeders have attempted to develop varieties with greater resistance to cold for expansion of blackberry culture into more northern areas. A recent introduction from the University of Illinois, 'Illini Hardy,' may represent a real breakthrough and opportunity for northern gardeners. The 'Illini Hardy' blackberry is an erect, thorny variety that is vigorous and very winter hardy. In field tests, 'Illini Hardy' was hardier, more vigorous, and produced fruit of more consistent quality than 'Darrow,' the standard erect thorny blackberry for the northern states. Tested at Urbana, Illinois, a site considered too cold for successful blackberry production, 'Illini Hardy' was the only surviving variety in test plots. This test included the winter of 1989-1990 in which the temperature dropped to −24°F (−31°C)!

There are considerably more blackberry variety options for home gardeners who live along the western and eastern seaboards and in the southern two-thirds of the U.S. In many cases, depending on your specific climatic region, you may be able to plant a combination of blackberry types that can extend the ripening season of these delicious treats for up to 10 weeks! In the central and lower piedmont North Carolina area, for example, winter temperatures seldom fall below 0°F (−18°C), and there the home gardener has the chance to produce blackberries from the end of the strawberry season through the middle of summer.

To get the blackberry season started, trailing blackberries (dewberries) will normally be the first to ripen. The trailing blackberries

were widely grown in the 1940's, but their soft fruit and requirement for staking or trellising led to their commercial demise in most
areas of the U.S., except in the Pacific Coast states where there are
still active dewberry and blackberry-raspberry hybrid growers.
Once important eastern dewberry varieties such as 'Lucretia' and
'Carolina' are becoming nearly impossible to find today. 'Marion'
trailing blackberry is one possible choice for gardeners in areas
where 'Lucretia' was once popular. The two major strengths of
dewberries are their earliness and good flavor.

The dewberry season is closely followed by that of the erect
thorny blackberries. 'Choctaw,' one of the newest releases developed by Dr. James N. Moore of the University of Arkansas Agricultural Experiment Station, is a very early erect thorny that will begin
ripening during the last week of May in North Carolina's lower
piedmont and coastal plain. The average ripening date of 'Choctaw'
in Clarksville, Arkansas is May 23. 'Choctaw' is to be feared,
however, for its "scratchy thorns." This variety will wound a friend
upon the slightest provocation, with no thought of an apology! Yet,
despite this fault, 'Choctaw' may deserve a place in your garden on
the basis of its earliness, good blackberry flavor, very small seed
size and rigid canes that do not require any external support (other
so-called "erect" thorny blackberries will benefit from some lateral
support for the canes during harvest).

In general, the erect thorny types of blackberries have excellent
fruit quality (generally much better than semi-trailing thornless varieties), but their thorns are understandably "objectionable" to
many home gardeners. However, due to the breeding success of Dr.
James N. Moore, there now exists two erect, *thornless* blackberry
varieties, 'Navaho' and 'Arapaho.' Prior to these Arkansas releases,
in 1988 and 1993 respectively, virtually all thornless blackberries
available in the U.S. had the semi-trailing growth habit, requiring a
trellis. The newest generation of thornless erect blackberries provides the best of both worlds, as they do not require any external
support for the canes, and berry quality is judged to be comparable
to that of the erect thorny blackberries. 'Arapaho,' the newest erect,
thornless blackberry, overcomes two important faults of 'Navaho':
it ripens earlier in the summer when temperatures in the South are
not as hot, and it produces new canes from its roots more reliably

than 'Navaho.' In other words, 'Navaho' is a very sparse producer of new "sucker" plants (arising from the roots), a characteristic that delays fruiting and row establishment. In terms of fruit ripening, 'Arapaho' is the earliest ripening thornless blackberry variety available at this time. 'Arapaho' will begin to ripen approximately 10 days after 'Choctaw.' Following 'Arapaho' in ripening is another relatively new erect thorny blackberry, 'Shawnee.' This variety is noted for consistent high fruit yields and very large fruit size. Some experts consider 'Shawnee' to be the best overall erect blackberry, and it surely deserves consideration equal to any other blackberry for home gardeners in the Midsouth and possibly along the Pacific Coast where the western trailing types are so popular ('Thornless Evergreen' blackberry, 'Marion' blackberry, 'Boysenberry,' 'Youngberry,' and 'Loganberry'). The western trailing types are high yielding, but the fruit bruise easily, and they are less cold hardy than the eastern erect blackberries.

The semi-trailing thornless blackberries may not qualify for a place in your home garden or landscape if you are unwilling to make a slight compromise on flavor. The lower dessert quality of the semi-trailing thornless blackberries may be excusable on the basis of their incredible productivity! Yields of 30 pounds per plant is not uncommon with an average semi-trailing thornless blackberry variety. Newer thornless varieties such as 'Chester' and 'Hull' are capable of producing crops exceeding 50 pounds per plant! These blackberries are especially desirable for jam, jellies, pies and juice. 'Chester' and 'Hull' also have reasonably good fresh eating quality. Semi-trailing thornless blackberries are typically a month later in ripening than the early to midseason erect thorny blackberries. The late ripening of the semi-trailing thornless blackberries may not be a drawback if you would like to have fresh blackberries from the garden well into July and even August. It is conceivable that a home garden consisting of trailing, erect and semi-trailing thornless types could have continuous production of blackberries for 8 to 10 weeks!

Two cautions about the semi-trailing thornless blackberries: First, these blackberries are not fully winter hardy, and they are best grown in the southern two-thirds of the U.S. In colder areas, such as New York, Pennsylvania, and Ohio, the canes and flower buds of semi-trailing thornless blackberries must be protected from winter injury by bending the canes to the ground in late fall before the ground freezes and covering the tips with soil. This keeps the canes below the snow line. Or, you can cover the canes with mulch. Using this method, Dr. Tomkins was able to grow 'Thornfree' blackberries at Ithaca, New York, and had a beautiful crop after a winter of − 18°F (− 28°C) during the coldest night! The second concern is related to avoiding the use of too much nitrogen fertilizer. This means no additional nitrogen fertilizer in late summer or fall for the semi-trailing thornless blackberries. If anything, many garden soils have too much accessible nitrogen, and the already vigorous semi-trailing thornless blackberry is much more disposed to winter injury if you stimulate fall-growth with fertilizer.

Success of a home blackberry patch depends largely upon selection of the proper varieties for your area. The next section provides some more specific information on popular blackberry varieties available today from small fruit nurseries in the U.S. Remember, a good variety selection for the Midsouth, such as 'Shawnee' erect thorny blackberry, may be a very poor selection for northern gardens because of susceptibility of the plants to winter injury. 'Darrow' is still the "standard" erect growing thorny blackberry for colder winter regions. But, the newer variety 'Illini Hardy' is showing great promise in these areas. It is always a good idea to cross reference the information provided here to the blackberry variety recommendations of your local Cooperative Extension Service county, regional or state center.

Trailing Blackberries (Require Staking or Trellis)

Hardy to 5°F to 10°F
and Have Medium Chilling Requirement

'Lucretia' was found in West Virginia in 1886, and was the leading cultivated dewberry for many years. It is sometimes called "bingleberry" and is one of the best cultivated dewberries for hardiness, large fruit size, and sweetness. The bluish fruit often reach 1 1/2 inches long. The plant requires a stake or trellis.

'Carolina' was developed in Raleigh, NC in 1951 and was the leading cultivated dewberry for many years in that area. 'Carolina' has resistance to cane and leaf diseases and excellent fruit sweetness. It requires a stake or trellis.

Erect, Thorny Blackberries (Require No Trellis)

Hardy to − 20° F to − 15° F (− 29° C to − 26° C)
and Have High Chilling Requirement

'Darrow' was developed and released in Geneva, New York in 1958. It is very erect; its fruit medium size, firm, good flavored and early ripening. The plants are vigorous and hardy to − 20°F (− 29°C) in the Northeast and Midwestern states.

'Illini Hardy' was developed in Urbana, Illinois and released in 1988. The fruit ripen late and have medium size and good flavor. The plants are vigorous and typically hardy to − 15°F (− 26°C) in the Northeast and Midwestern states. Some nurseries claim that 'Illini Hardy' will produce a more consistent quality fruit than 'Darrow.' Plants are resistant to Phytophthora root rot.

Hardy to − 10° F (− 23° C)
and Have Medium to High Chilling Requirement

'Arapaho' was developed in Arkansas and released in 1993. This erect-caned, thornless blackberry does not require trellis support. 'Arapaho' fruit are early ripening (11 days before 'Navaho') and the plant produces new canes from roots easily for quick establishment of a full fruiting hedgerow. The fruit are firm, short, conic and bright glossy black. Flavor is equal to 'Navaho' and better than 'Shawnee' and 'Choctaw.' The plants appear to be immune to rosette and orange rust. Another positive characteristic of 'Arapaho' is its small seed size. This variety is adapted to the Southeastern and Midwestern U.S.

'Cherokee' was developed in Arkansas and released in 1974. The plants have very erect canes that require no trellis or lateral support. 'Cherokee' fruit ripen early (same season as 'Cheyenne') and have very good flavor. The berries are medium-small in size. The plant is resistant to anthracnose but susceptible to rosette, and is adapted to the Southeastern and Midwestern U.S.

'Cheyenne' was developed in Arkansas and released in 1976. Its canes are erect, but the plants will benefit from lateral support at fruiting. 'Cheyenne' fruit ripen early and are among the firmest erect thorny blackberries. Fruit size is medium. 'Cheyenne' has good flavor and is excellent for processing. The plants are resistant to orange rust, but susceptible to rosette, and are adapted to the Southeastern and Midwestern U.S.

'Shawnee' was developed in Arkansas and released in 1985. The very large fruit ripen late but have excellent flavor. The plant is very productive over a 3 to 4 week period. Many growers consider 'Shawnee' to be the best erect thorny blackberry for the home garden. Along with 'Cheyenne,' 'Shawnee' canes benefit from lateral support at harvest time. The fruit is less firm than those of 'Cheyenne' but more firm than those of 'Choctaw.' 'Shawnee' is resistant to orange rust, but susceptible to rosette. This variety is adapted to the Southeastern and Midwestern U.S.

Hardy to − 5°F (− 21°C)
and Have Medium Chilling Requirement

'Choctaw' was developed in Arkansas and released in 1988. It is the earliest ripening erect blackberry (thorny or thornless). The fruit are medium in size and have excellent flavor. Seed size is small. The plants are high yielding and resistant to orange rust but susceptible to rosette. They are adapted to the Midsouth.

'Navaho' was developed in Arkansas and released in 1988. This was the first erect, thornless blackberry to require no trellis support. Late ripening of the fruit may be considered a drawback in the South where summer temperatures are often too hot in late June and July. 'Navaho' does not sprout well from root cuttings and I recommend you purchase only rooted plants of this variety. The fruit are medium in size, have excellent flavor, and are the firmest of all erect blackberry fruit. Seed size is similar to that of 'Shawnee' and smaller than that of other thornless semi-trailing blackberries. Homeowners who strongly object to blackberry thorns or prickles may wish to grow 'Arapaho' (early) or 'Navaho' (late) for an extended ripening period. 'Navaho' has shown good resistance to rosette, but occasionally the plant is infected with orange rust. This plant is adapted to the Midsouth.

Hardy to 0° to 5°F (− 18° to − 21°C)
and Medium to Low Chilling Requirements

'Brazos' was developed in Texas and released in 1959. The fruit ripen very early and are large and acid, and the seeds are large. The plant is moderately susceptible to rosette, but has become a standard variety in Texas and along the Gulf Coast, and can be grown north to central Arkansas. They are hardy to 0°F (− 18°C).

'Humble' was developed in Texas and released in 1942. The small, soft fruit ripen early to midseason. The plant has medium vigor, is not high yielding, but it is resistant to rosette. It is also susceptible to orange rust. This variety is adapted to Texas and the Gulf Coast, and may be hardy to − 5°F (− 21°C).

FLORIDA AND GULF COAST REGION

'Flordagrand' was developed in Florida and released in 1958. The very early fruit are large and soft, but have good flavor. The plants are vigorous and productive and require cross-pollination. This variety has a short chill requirement.

'Gem' was developed in Georgia and released in 1967. The fruit are large, firm, good-flavored and ripen early. The plant is vigorous, productive, and thorny.

'Oklawaha' was developed in Florida and released in 1964. The fruit ripen very early, are medium-sized and soft, but good flavored. The plants are vigorous, semi-evergreen, and have a short chilling requirement.

'Young' was developed in Louisiana and released in 1926. The fruit are very large, sweet, wine colored, and ripen midseason. The plants are vigorous and very productive, but susceptible to rosette. They are adapted to the Gulf Coast and California.

Semi-Trailing, Thornless Blackberries
(Late Ripening Unless Noted Otherwise)

'Black Satin' was developed in Illinois and released in 1974. The semi-trailing, thornless canes are vigorous and productive. The fruit are large, firm, tart, and ripen very late. The plants are adapted to the Midwestern and Middle Atlantic States.

'Dirksen' was developed in Illinois and released in 1974. The semi-trailing, thornless canes require trellis support. 'Dirksen' is an earlier blooming and earlier ripening thornless that is hardier than 'Black Satin' in higher elevation mountainous areas. Its fruit have good flavor. 'Dirksen' has less vigor than 'Black Satin,' 'Hull' and 'Chester' and yields decline as the planting ages. It is adapted to the Midwestern and Middle Atlantic States and to the Midsouth.

'Georgia Thornless' was developed in Georgia and released in 1967. This is a thornless variety that ripens in the erect blackberry mid-season. The fruit is large and good flavored. The plants are productive and adapted to the Gulf Coast region. They are not hardy below $10°F$ ($-12°C$).

'Hull' was developed by the USDA at Beltsville, Maryland and released in 1981. It has semi-erect, thornless canes that require trellis support. The fruit are medium large and somewhat acid if picked before they are a fully ripe dull black. Fully ripe fruit have excellent flavor. The plants' susceptibility to rosette is not known. A very long ripening season makes it especially suitable for the home garden, or to extend the commercial harvest of blackberries into the summer months. This was meant to be a superior replacement for 'Black Satin' and is adapted to the Middle Atlantic, Midsouth, and Midwestern states.

'Chester' is a thornless, semi-erect variety released by the USDA in 1985. It ripens about seven to 10 days after 'Hull.' The fruit are somewhat acid if picked before they are a fully ripe dull black, but when fully ripe have an excellent flavor. Fruit size is similar to that of its sister seedling 'Hull.' Trellis support is required. The plant is adapted to the Midwest, Middle Atlantic, and Midsouth regions.

SITE AND SOIL CONSIDERATIONS

The blackberry hunter finds wild berries in places that are often partially shaded by trees, though this is seldom the best place to grow them. If there is any choice within the yard, choose the sunniest site possible for blackberries. Two other site considerations for blackberries are air drainage and water drainage. Blackberries, especially the trailing types and earliest ripening erect blackberries such as 'Choctaw,' are potentially subject to damage from spring frosts at bloom time. Elevated sites with good cold air drainage provide a degree of frost protection and are preferable but not an absolute necessity. Erect blackberry varieties generally bloom late (April) and are seldom damaged by spring frosts.

Blackberries are damaged when water stands around their roots at any time of the year, and poorly drained soils should be modified or avoided.

Windbreaks are recommended to prevent cane damage in open, windy areas. Orientating erect blackberry rows so that they run in the direction of strong prevailing winds is beneficial, since blackberry plants tend to support each other and this minimizes cane breakage. It is important to avoid cross winds. However, some amount of air movement through the blackberry patch is actually very desirable. Good air circulation reduces high humidity around the canes. High humidity favors the development of several serious blackberry diseases, especially double blossom or rosette. Thus, sites with good air drainage and rapid drying are likely to be less troubled by disease. The direction of the site slope is not a critical factor in selection, but it can affect the time of bloom and fruit ripening. Blackberries on a southern slope will usually bloom two to three days earlier than those on a northern slope.

The availability of water for irrigation near the planting is a real advantage. Blackberries require water, especially when fruit is sizing up and ripening. At least 1 inch of rainfall per week is necessary to obtain optimum fruit size and yield. Rainfall may be supplemented with irrigation—either drip or overhead sprinklers.

Another factor worthy of consideration is the previous cropping history of a potential site *AND* its proximity to wild brambles. Certain pathogens and diseases can build up in the soil where certain crops were previously planted. Do not plant blackberries immediately following potatoes, tomatoes, peppers, or eggplant, since this increases the risk of infection with verticillium wilt. Avoid a site previously planted with fruit crops such as peaches, apples, grapes, or brambles because of the potential for crown gall infection. The Arkansas thorny erect blackberries are susceptible to double blossom disease and should not be planted within 300 feet of wild blackberries.

Test the soil the season before planting, and apply fertilizer and lime according to recommendations. Blackberries appear to do better in more acid soils than most other small fruits, except acid-loving blueberries. Soil pH above 7.5 can cause iron chlorosis problems such as those observed on some of the 'chalky' high pH soils in Texas.

CULTURE

Site Preparation and Planting

Prepare your site well in advance of planting, as described in the Introductory chapter.

Purchase certified planting stock well in advance of planting.

Erect Blackberries

Root cuttings are the preferred planting material (except for the upright, thornless blackberry 'Navaho'). Rooted plants (suckers) are also easily dug and transplanted. Dig root cuttings in January or February in the Midsouth, February or March in the Middle Atlantic and Midwestern states, and March to early April in the Northeastern states, from healthy, disease-free plantings. Roots pencil-size to one-half inch in diameter should be cut to four to six inches in length (Figure 1) and protected from drying. Store them, until planting, in sealed polyethylene bags in above-freezing refrigeration, or by bundling and burying them in moist soil.

Plant erect blackberries in late winter—February or early March is ideal in most of the Southeastern states. April or early May planting is desirable in the Middle Atlantic and Midwestern states, and early to mid May in the Northeastern states. Research in Arkansas has shown that highest productivity in that region can be attained if they are planted before April. Of course, if the soil is wet, delay planting until conditions improve. The growth of blackberry plants set late in the spring can be seriously checked by drought or drying winds.

FIGURE 1. Healthy, disease-free .5 inch diameter blackberry roots cut into 4- to 6-inch lengths.

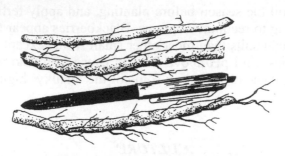

Blackberry root pieces are dropped "horizontally" about two feet apart in a shallow furrow, and covered with two to three inches of fine top soil. On tighter clay soils, cover the root pieces with only one or two inches of soil. *Never* place fertilizer in the furrow with the blackberry root cuttings. Rooted plants (suckers) are set about 3 feet apart in the row. Space rows of erect blackberries 10 feet apart.

New leafy shoots from the root pieces will emerge erratically in the spring after the soil has warmed. An important caution: the young plants are extremely vulnerable to weed competition, and hand hoeing is the most practical control. Preemergent herbicides are applied in commercial plantings, but I strongly recommend the use of "iron oxide" (hoeing) in the home garden.

Trailing and Semi-Trailing Blackberries

The semi-trailing thornless plants do not naturally propagate by suckers or from root-cuttings as do the erect blackberries. Trailing blackberries (dewberries) are sometimes propagated from root cuttings, but trailing and semi-trailing blackberries (like black raspberries) are mainly propagated by tip-rooting (tip-layers) (Figure 2). Tip layered plants are usually available from November to March. Late fall planting is possible, provided the soil is in suitable condition. However, early spring planting of dormant tip-layers is generally best.

If the plants are dry upon arrival, soak the roots in water for several hours before planting or bedding in. If you do not plant

FIGURE 2. Tip-layered blackberry plants. 2a. New plants are propagated by tip-rooting in August. 2b. A hole 6 inches deep is dug near the parent plant, and the tip of a young shoot is bent down into it. 2c. The following early spring, the rooted tip is cut from the parent plant with a 10 inch cane, and then dug up and planted in a new location.

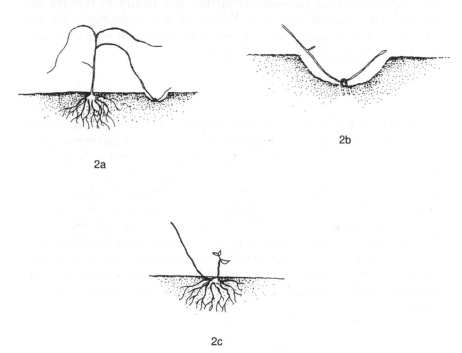

2a

2b

2c

immediately, heel in the plants by digging a trench deep enough to contain the roots. Spread the plants along the trench, roots down, and cover the roots with moist soil. Be sure to keep the ground moist.

Semi-trailing thornless tip-layer plants are quite vigorous and should be set 6 to 10 feet apart in rows. Less vigorous trailing types, like 'Carolina' dewberry, are set 4 feet apart in rows 8 feet wide for wire trellis, and 6 feet wide for stakes. Align plants carefully in the row to accommodate the trellis, which will be constructed for semi-trailing blackberries.

Use a hand trowel or fork to dig a hole wide enough and deep enough to allow the roots to spread out naturally. Spread the roots out in the planting hole. Planting depth should be slightly deeper than the tip layer or rooted plant (erect blackberry grown in the nursery) to permit some settling of the soil around the base of the plant (crown). After planting, tamp the soil firmly to remove air pockets from around the roots. Water all new plantings well immediately after planting. Also, at this time, cut down each cane to a bud about 10 inches above the ground.

Fertility the First Year

The type and quantity of fertilizer applied before planting should be based on soil test recommendations. In good blackberry soils, nitrogen is usually the only seriously limiting nutrient. Needs for phosphorus, potassium and other elements should have been corrected in the soil preparation period before planting. If plants fail to initiate vigorous growth, an application of 1 pound of 10-10-10 fertilizer or equivalent per 20 feet of row can be applied in the spring of the first year. Avoid using nitrogen fertilizer later than July since this may result in late growth and subsequent winter injury. If plants are not fertilized or irrigated after mid-summer, their dormant canes and fruit buds will be more resistant to low winter temperatures.

FIRST YEAR CULTIVATION AND TRAINING

A key consideration following planting should be control of weeds. Blackberries may be shallowly cultivated during the first growing season, but care must be taken in erect blackberries to prevent breaking the tender, newly emerging plants arising from root pieces. In addition, in cultivating both erect blackberries and trailing and semi-trailing types in early spring, you must be "extra careful" not to hoe up to the base of newly set plants as you may injure new shoots that are trying to emerge from what are called "crown buds." These crown buds, or "leader buds," are located at the base of the overwintering canes, and generally out of sight and lightly covered with soil after transplanting. Both the trailing and semi-trailing blackberries produce all of their new shoots (called primocanes) from crown buds or leader buds at the base of the original transplant. Erect blackberries produce new shoots from both crown buds and from "sucker shoots" coming from the roots. Once your original erect blackberry plants become established in the spring, they proceed to develop a network of fibrous roots that explore the surface soil area in your garden. In the late spring and early summer of the first year, the blackberry root system network will produce shoots at random locations "away" from the original set plant. Basically, erect type blackberries are grown in hedgerows, wherein the sucker plants arising from the blackberry roots fill in the entire row space (Figure 3). The base of the row should be kept to a width of about 1 1/2 feet by removing or hoeing out suckers that arise beyond this limit. The best plant density is 4 to 6 vigorous plants per lineal foot of blackberry hedgerow. Do not be overly

FIGURE 3. Erect type blackberry plants produce new shoots from the crown of the original set plant (crown bud) and the buds formed on the roots. Topping of new shoots in early and midsummer will force lateral branching. Blackberries are perennial plants with a biennial growth and fruiting habit. The perennial part is in a storage root. The biennial part is the new growth (primocanes) which can overwinter, flower (floricanes) and fruit in the second season, and die after fruiting.

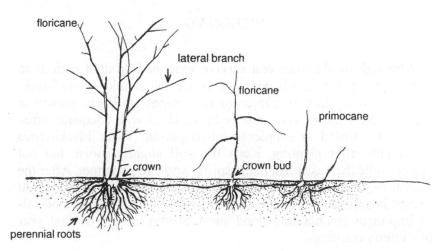

concerned if you do not achieve this density in the first growing season–your primary goal in the first year is to encourage good growth and development in the plant material you set in the early spring. I advise against trying to produce any fruit on the canes produced in the first growing season. It is best to simply cut and remove all low and sprawling first year canes of erect blackberries in February or March of the second year–make your cuts as close to the ground as possible, and remove these canes from the garden area and have them buried or burned (if local ordinances will permit burning).

Semi-trailing blackberries do not spread by underground root suckers–their new growth is confined to the plant crown. First year training of the shoots arising from crown buds is fairly simple–let them sprawl on the ground during the growing season, and then cut them completely OFF in February or March of the second year (same procedure outlined for erect blackberries). In general, first year growth of any type of blackberry is low and sprawling–even the erect types. The fruit yield potential is so poor in the second year that it is definitely not worth the effort to train the shoots of trailing and semi-trailing blackberries to a trellis in the first growing season.

WATERING

Although blackberries can survive extended drought periods in their native habitats, this does not mean you should ignore blackberries growing in your garden or landscape! Optimum growth in the first growing season can only be realized with adequate moisture from rainfall or supplemental irrigation. Water blackberries thoroughly after planting. Keep the soil slightly moist, but not soggy while they establish in the spring and early summer. After the month of June, additional watering is generally not as important except in periods of drought. In the Midsouth, irrigation is especially important in obtaining good survival and growth of first year blackberry plantings.

PRUNING AND TRAINING

Growth Habit

Blackberry plants produce new shoots from the crown of the original set plant and possibly from buds formed on the roots (sucker plants). These shoots, called 'primocanes,' grow one season and produce laterals (side branches). The second year, small branches grow from the buds on the laterals. Fruit is borne on the tips of these small branches. After the two-year canes have borne a crop, the cane dies (the canes are biennial). As soon as the last berries have been picked, the old canes that have just borne fruit should be cut and removed. This allows the young canes more room for development, conserves the moisture supply, and destroys any insects or diseases on the old canes. Understanding the blackberry growth habit is necessary for proper training and pruning.

Erect Blackberries

Erect blackberries such as 'Cherokee' and 'Cheyenne' send up root suckers in addition to new shoots that arise from the crown. If all root suckers were allowed to grow, they would soon turn the blackberry plantation into a thicket. During the growing season, it is desirable to allow the suckers to develop in a row approximately 1 1/2 feet wide, but it is very important to pull out suckers that grow outside the row.

When the new shoots of erect blackberries reach a height of 30 to 36 inches, cut off the tips. This makes the canes branch and provides a greater bearing surface for high yields. Tipped canes also grow stout and are better able to support a heavy fruit crop than untipped canes, which must be supported by a trellis.

In winter, prune the laterals to 12 to 16 inches for convenient harvesting and larger berries. In late winter, remove the remaining dead and weak wood. Leave healthy, vigorous canes spaced about 6 canes per lineal foot of row.

In summer, as soon as the last berries have been picked, cut out all the old canes and burn them. Also, you should top new shoots at this time, in order to force lateral branching.

Semi-Trailing Thornless Blackberries

During the first growing season, thornless blackberry shoots (a shoot is the current season's top growth) will tend to have a trailing habit, and shoots are generally left on the ground. Then, before bud swell in the second season, bring the canes (a cane is a mature shoot after it has lost its leaves) up to the trellis wires and tie them individually with soft string. The lateral branches are pruned to 10 to 12 inches.

Often, only a small crop is available for harvest the year after planting. For this reason, some growers cut back to within several inches of the ground canes that would have otherwise fruited. This helps the plants become better established by preventing a severe drain on their vitality from fruiting, and by reducing heat and drought stress. It also favors the development of sturdier, more fruitful shoots.

In the second and succeeding years, new shoot growth is more vigorous and upright. These shoots should be tied to the trellis as soon as they have reached a height of 4 to 5 feet. Fan the canes out from the ground and tie them where they cross each wire. *Avoid tying canes in bundles.*

Prune in late winter or early spring. Laterals on canes should be pruned back to a length of 12 to 18 inches. Fruit from pruned laterals is larger and of better market quality than fruit from unpruned laterals.

In summer, as soon as the last berries have been picked, cut out all the old canes. *Do not* remove the new canes which have come up since spring, except to thin to 4 to 8 shoots per crown. The best shoots should be selected so that wires are well covered with evenly spaced shoots. Broken shoots or those too short or too weak for training should be removed.

Ordinarily, no further summer pruning is performed on thornless semi-trailing blackberry varieties. However, research in Maryland has indicated potential benefits from periodic summer topping to encourage more lateral branching and the development of shorter, more compact plants. Plants should be set closer than six feet in the row you follow if this management plan.

Several innovative researchers have used a two sided trellis. In this system, the fruiting canes are tied to one side of a trellis resembling a Geneva Double Curtain in grape production. As current season shoots develop they are trained to the opposite side of the trellis. This system gives a V-shaped row. After harvest, old canes on the fruiting side of the row are mechanically pruned near the ground without much injury to current season shoots which have been tied to the other side of the trellis. The two-sided trellis shows promise of being very productive.

A more detailed year-by-year schedule for home gardeners interested in a scientific (and somewhat complicated) approach to semi-trailing thornless blackberries is given in the Appendix at the end of this chapter.

Trellis Support

After the first season, train trailing and semi-trailing thornless blackberries to trellises to assure clean fruit, easy picking, and easier disease control.

Many trellis arrangements and training methods are satisfactory. To construct a simple and effective trellis: stretch two wires (gauge size 9 or 11) between line posts set 18 to 24 feet apart in the row. String one wire three feet from the ground and the other about five feet from the ground (Figure 4). Wires should be stapled loosely enough to allow for contraction in cold weather.

Trellises of this height require sturdy end posts, 8 feet in length, well braced, and anchored. Line posts should be about 7 feet long and 3 inches in diameter.

Trailing Blackberries

Trailing blackberries need support to keep the fruit clean and to make harvesting easier. The system of training to be used depends on the climate, the cost of materials, and individual preference.

In North Carolina, where the 'Lucretia' variety is almost the only variety grown commercially, the plants are usually set 5 by 5 feet and the stakes are usually at least 5 feet high. Under this system cross-cultivation is possible and less hand labor is needed.

In North Carolina and other states having a similar or a longer growing season, all the canes, both old and new, are cut off close to the crown of the plant immediately after harvest to aid in controlling rosette (double blossom), leaf spot, and anthracnose. Because

FIGURE 4. 4a. Canes of semi-trailing blackberries trained along two wires. Semi-trailing thornless blackberries require sturdy end posts, 8 feet in length, well braced and anchored. 4b. Alternatively, a clothesline support system can be used to advantage for higher yields with semi-trailing black-berries.

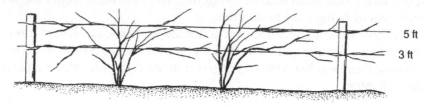

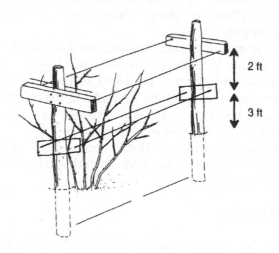

the next year's crop depends on the amount of vigorous new growth, this cutting out must be done promptly. Figure 5 shows a tool used for this purpose. Apply fertilizer and begin cultivation at once and continue until the new canes begin to interfere. The new canes are allowed to run on the ground and are left there until next spring, when they are tied to stakes in the same manner as the previous year's growth.

A wire trellis is used throughout the U.S. for the 'Young' and

FIGURE 5. Long-handled shears used in North Carolina for pruning trailing blackberries. Note that the steel blades are angled upward, thus enabling the pruner to cut the canes close to the crown without much stooping.

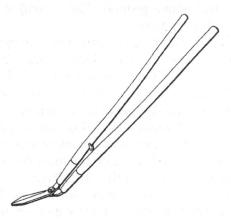

'Boysen' varieties and in southwestern Michigan, where stakes are expensive, for the 'Lucretia.' The 'Lucretia' plants are set about 2 to 2.5 feet apart in rows 7 to 8 feet apart. Posts are set along the row from 20 to 40 feet apart, depending on their strength and the vigor of the canes. The posts project about 3.5 feet above ground, and a wire is stretched along them about 2.5 feet above ground. The canes are gathered together in a bunch and tied to the wire. The ends are usually cut off 8 to 12 inches above the wire. This system is the principal one used in southwestern Michigan, and it is sometimes used in eastern New York. Sometimes, however, the ends are tied along the wire and left to bear fruit. The two-wire trellis is liked by many growers for the 'Young' and 'Boysen' varieties.

In Kansas sometimes no trellis is used. The canes are trained along the ground in a row 2 feet wide and are lightly covered with straw in late fall. This straw is left on during the next spring, and the fruiting laterals grow up through the straw so that picking is much like that of strawberries.

In Pacific coast plantings, very little staking is done, but a few growers use cedar stakes set 6 to 8 feet apart each way and 6.5 to 7 feet above ground. Sometimes a cross arm is used, and the canes are looped over it.

In California, Oregon, and Washington, the two-wire vertical trellis is most commonly used but often only a single wire is used. Both the single wire and the top wire of the two-wire trellis are about 4.5 to 5.5 feet above ground. The second wire, if used, is about 2 feet below the top wire.

During the summer, Pacific coast growers train the new canes on the ground along the row. They are held in place by small stakes about 2 feet long or by stiff wires curved into a U-shape and placed over the canes. In late summer or more commonly in early spring, the canes are trained in various ways on the wires of the trellis.

The oldest method of training is that of the wreath, or weaving, system, in which one or a few canes are taken at a time and passed from upper to lower wires. Presently, the most common method is to divide the canes of a single hill, raise them to the upper wire, make one or two twists, and bring them down to the lower wire and back to the plant. Then the ends of the canes are usually cut off. Sometimes tying may be necessary to hold the canes in place. This if often called the loop method of training (Figure 6). Often the canes are divided into 2, 3 or 4 bundles as in Figure 7 and tied to a single wire.

Fertility Management

Any fertilizer needs for the initial growing season should have been identified by soil tests and corrected during preplant land preparation. However, if the plants fail to initiate vigorous growth, additional nutrients can be applied in the spring. Avoid use of nitrogen fertilizer later than July since this may result in subsequent winter injury. Fertilizer suggestions (Tables 2 and 3) are presented for your reference.

Water Management

Lack of water just prior to or during the harvest season can seriously reduce the blackberry crop. A shortage of water at these times will not only affect the current season's crops, but will also limit the production of desirable fruiting canes and so affect the following year's crop.

FIGURE 6. Loop method of training 'Logan' and other trailing blackberries to (a) a 2-wire or (b) a 3-wire trellis. Note that the canes are arched over the top wire and then looped around the second one.

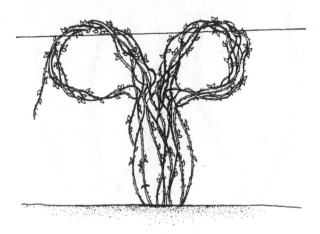

a

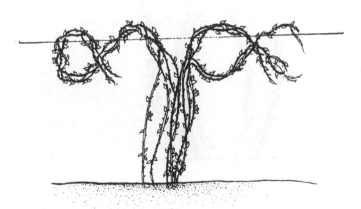

b

FIGURE 7. Trailing blackberry canes can also be divided into two (7a), three (7b) or four (7c) bundles.

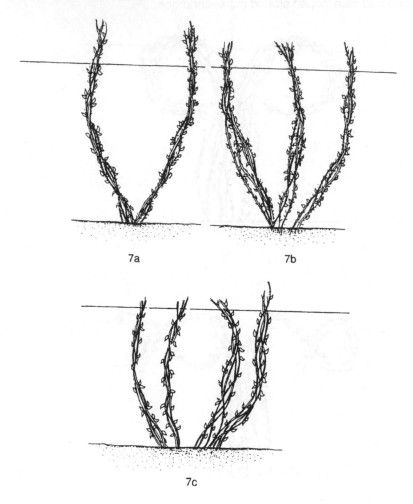

7a

7b

7c

It is generally recommended that plants receive at least one inch of water during each seven to ten day interval during the growing season. The amount of water can be reduced, perhaps by 50 to 60 percent, if trickle or a similar irrigation system is employed, since only the soil around the plant is wet. The lower-volume trickle method is especially well-suited to semi-trailing thornless blackber-

TABLE 2. Fertilizer recommendations for erect thorny blackberries.

Time to apply	Method of application	Kind of fertilizer	Amount to apply per acre
Planting	None	None	None
Just after growth starts in spring	Banded at the side of the row (6 inches from row)[1]	10-10-10	approximately 5 lbs/100 ft row
Second year every year thereafter	Broadcast fertilizer alongside row in late winter/early spring	10-10-10	approximately 5 lbs/100 ft row
	Sidedress after harvest to provide virgorous new canes	Ammonium nitrate	approximately 2 1/2 lbs/100 ft row

[1] Do not apply fertilizer in the furrow in which root cuttings or plants are placed.

TABLE 3. Fertilizer recommendations for semi-trailing thornless blackberries.

Time to apply	Method of application	Kind of fertilizer	Amount to apply per acre
Planting	None	None	None
Just after growth starts in spring	Spread uniformly in 4 inch bands around, but not closer than 6 inches from transplant	10-10-10	3-4 oz/transplant
Second year and every year thereafter in March	Spread uniformly in a 3 ft wide band over the row or sidedress with 1/2 recommended on each side of the row. Sidedress bands should be 1 ft wide and 16-18 inches from row-center.	Ammonium nitrate[1]	3-7 lbs/100 ft row

[1] The soil should be tested every 2 years for acidity, phosphorous, potassium, calcium and magnesium. If these tests indicate a pH of 5.7 or higher, and medium or high levels of the above nutrients, only nitrogen is needed on an annual basis. Lower pH readings, or low levels of individual nutrients, indicate a need for lime and/or a more complete fertilizer such as 10-10-10.

ries where new growth is confined to the immediate area of the original plant crown (these blackberries do not spread by underground root suckers). The trickle irrigation method will also reduce the risks of fruit rot by avoiding the need to wet the foliage and fruit during water application. However, the trickle system will rapidly become clogged if the water contains impurities. Both well water and surface water from ponds or streams must be tested, and the required treatment and filtration provided for the system to function properly.

Mulches can be used but are often costly and scarce. Large volumes of material and labor are required, and some must be replaced each year. Mulches may introduce weed seeds, encourage rodent infestation and crown gall, and can be a fire hazard. However, weed free mulches, such as straw, pine needles, sawdust, wood chips, or other suitable materials, are of much value in aiding soil moisture retention and adding organic matter to the soil. They should be given serious consideration if you are growing blackberries on lighter, low organic matter soils without supplemental irrigation.

Weeding

The first consideration following planting should be effective weed control. Blackberries may be shallowly cultivated during the first growing season, but take care to prevent breaking the tender, newly emerging primocanes. Generally, blackberry plantings are clean-cultivated between the rows by shallow tilling. As the planting ages, blackberry roots will invade the area between the rows, and cultivating too deeply will injure roots and induce unwanted suckering between the rows. If sod is allowed to develop between rows, it should be mowed several times during the growing season. To avoid soil erosion, establish a grass sod between rows. If sod is used in the middles, a 4-foot wide grass-free, weed-free strip must be kept in the plant row.

Insects and Diseases

Many insects and diseases can damage blackberries. However, only a few of these problems are likely to occur in a given area.

Damage from insects and diseases can be kept to a minimum if these four general suggestions are followed: (1) Remove all wild blackberry and raspberry plants in the vicinity of the field. (2) Select high quality planting stock. Purchase certified stock whenever possible. Be cautious about accepting plants from neighbors. (3) Destroy plants in which diseases appear, and prune out insect infested canes and burn or bury them away from the planting. (4) Remove old canes after harvest from the field.

If a problem arises, have it identified by your County Extension Center or an employee of a good garden store. More information on insect and disease problems may be found in publications obtained from your local County Extension Center.

Harvesting

Harvest blackberries at least twice each week, but do not pick blackberries as soon as they turn black. It's better to wait three or four days and pick when the color appears dull and the berry separates easily from the plant. Some general guidelines for harvesting blackberries are: (1) Pick in the morning while the temperature is still cool and the berries are firm. (2) Pick and handle the fruit carefully to avoid crushing or bruising. (3) Pick berries when fully ripened (dull black appearance). (4) Gently place the berries no more than two inches deep in berry baskets or picking containers to avoid further bruising. (5) Cool the fruit as soon as possible after harvest.

SUGGESTED READINGS

Moore, J.N. 1990. Blackberry management. pp. 214-244. In G.J. Galletta and D.G. Himelrick (eds.). *Small Fruit Crop Management.* Prentice Hall, Englewood Cliffs, NJ.

Jennings, D.L. 1988. *Raspberries and Blackberries: Their Breeding, Diseases and Growth.* Academic Press Inc., San Diego, CA.

Tomkins, J.P. 1977. Cane and bush fruits are the berries; often it's grow them or go without. pp. 272-278. In Growing Fruits and Nuts. U.S.D.A., Agric. Info. Bulletin No. 408.

APPENDIX

*Advanced Pruning and Training Techniques
for Semi-Trailing Blackberries*

First Year

1. Remove the above-ground portions of the canes (the handle) after planting.
2. Do not summer prune: avoid damage to developing shoots as these may, when vigorous, provide a crop in the following year.
3. If weed control is poor or if sod middles have been established, tie new shoots to the trellis as soon as they are long enough to reach the bottom wire.
4. During the late dormant period (i.e., just before buds begin to swell in the spring), retain 2 or 3 of the most vigorous canes (remove all others), train and tie these shoots to the trellis in a uniform "fan" pattern. Head canes and laterals back to remove dead or small diameter (less than .25 inch) wood and shorten retained canes to 7 or 8 feet in length for the first fruiting season.

Second Year (includes first fruiting season)

1. Strip buds from the lower part (i.e., within 18 to 20 inches of the soil) of each retained cane after buds have swollen and produced .5 to 1 inches of new growth. This will eliminate fruiting shoots whose berries would otherwise have contacted the soil and become unusable. Stripping buds from the lower canes may be omitted where labor supply or costs are limiting factors, but this may affect plant capacity to maximize yields of usable fruit.
2. Remove or sever shoots (canes) which were retained in dormant pruning, as soon as harvest has been completed for the year: usually in mid- to late August, depending upon the cultivar.
3. Summer prune by removing 10 to 12 inches from tips of new main shoots after they have become long enough to tie in place on the top wire of the trellis. Generally, cuts should be 4 to 6 inches above the top wire after tying. There may be increased danger of cane blight if summer pruning is done too soon before, during, or after periods of rainfall or irrigation. Removal of shoot-tips may not be needed in some training systems.
4. Prevent rooting of shoot tips, during late summer and fall, by tying shoots up.

5. During late dormancy, select and retain 3 to 8 of the most vigorous main canes produced during the preceding year for production of fruit in the following summer. Adjust the number of canes per plant according to the amount of growth which occurred during the previous summer. Train canes in a fan pattern away from the crown and place ties where canes cross each trellis wire. Lateral shoots may be shortened to lengths of 10 to 20 inches; or if they are vigorous and originate below the bottom wire, they may be pruned longer and tied to the upper trellis wires as though they were main canes. Head-back (i.e., shorten) main canes as necessary to prevent competition among plants, but allow shoots of adjacent plants to overlap a foot or two at their ends. Total retained cane lengths (main canes plus laterals) of 20 to 50 feet are probably appropriate, depending upon plant size or vigor, for this age plant (Figure 8).

Third Year and After

1. Follow the same pruning procedures as for the second year (above), but the growth habit of these cultivars changes as plants and crowns become larger and more mature. There may be a relatively small number of very large diameter canes and few or no small diameter canes originating from a given crown; in this case, the laterals may bear almost all the fruit. Care must be taken to ensure that sufficient laterals are retained during dormant pruning to total a length of approximately 55 feet (e.g., 37 laterals averaging 18 inches in length). The suggestion for leaving 55 feet of laterals is based upon preliminary observations and it is subject to change as more information is accumulated.

2. Plants which bore heavy fruit loads while producing numerous canes or long laterals during the previous year should be pruned to the highest number and length of retained canes. Total cane lengths of approximately 100 feet per plant are suggested for large, highly vigorous plants.

3. Plants that have low vigor should be pruned to retain fewer canes. Low vigor may indicate that pruning was not severe enough in the previous year. Low vigor may also be caused by insufficient rainfall and irrigation, insufficient nitrogen (or other) fertilization, or by the effects of insects (esp. red raspberry crown borer) and diseases.

4. Excessive vigor may be the result of pruning which was too severe during the previous season, too much nitrogen, or too much rainfall or irrigation. Excess vigor also may result when winter injury reduces or eliminates cropping in the previous year. In the latter case, be cautious in adjusting your pruning severity; such vigorous vegetative growth is not likely to recur when a full crop load is present on the plant.

FIGURE 8. Training systems. Two systems are commonly employed: (A) Fan system—the fruiting canes are tied singly to the wires from a fan filling the whole space, except immediately in the center where space is left for developing new shoots (they may temporarily be tied to a 6 foot wire). This method is costly because of time taken to do the tying. (B) Horizontal or Arm system— tie canes horizontally along the wires similar to a Kniffen pruning system used on bunch grapes. The fruiting canes may all be tied to the wires on one side of the crown, and the new shoots similarly tied to the other side to make postharvest pruning operations easier. This system uses less labor, but the crop is generally as heavy, particularly where the fruiting canes are tied to one side.

A

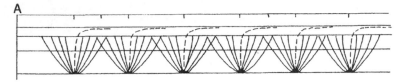

B

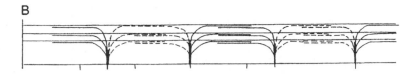

Blueberries–
North and South

R.E. Gough

INTRODUCTION

The blueberry is also known variously, and sometimes incorrectly, as whortleberry, hurtleberry, sparkleberry, bilberry, and huckleberry. It became established in North America after the latest ice age and its fruit were regarded highly by Native Americans who added them liberally to puddings, cakes, and pemmican. Some fruit were preserved by drying in the sun and the leaves, when dried and chewed, yielded a mild narcotic. Early American settlers also valued the berries and used the strong, flexible wood for tool handles.

Native Americans knew that the blueberry grew best in open areas and periodically burned the forests to encourage the spread and productivity of these useful plants. Burning also encouraged new growth from other plants which attracted browsing animals to be used for game. In fact, there are a number of native dishes that include blueberries and game in intimate combinations.

Blueberries are one of the easiest fruit to grow and once established, are virtually impossible to kill. In clearing ground for a new house, I once bulldozed a patch of wild blueberries with a John

R.E. Gough is Associate Professor and Extension Horticulture Specialist, Department of Plant, Soil, and Environmental Sciences, Montana State University and Editor, *Journal of Small Fruit & Viticulture*.

[Haworth co-indexing entry note]: "Blueberries–North and South." Gough, R.E. Co-published simultaneously in *Journal of Small Fruit & Viticulture* (Food Products Press, an imprint of The Haworth Press, Inc.) Vol. 4, No. 1/2, 1996, pp. 71-106; and: *Small Fruits in the Home Garden* (ed: Robert E. Gough, and E. Barclay Poling) Food Products Press, an imprint of The Haworth Press, Inc., 1996, pp. 71-106. Single or multiple copies of this article are available from The Haworth Document Delivery Service [1-800-342-9678, 9:00 a.m. - 5:00 p.m. (EST). E-mail address: getinfo@haworth.com].

Deere 450. The following spring the area was dense with new blueberry shoots arising from the old crowns and rhizomes.

Fortunately, today we don't have to depend upon wild plants for fruit; plant breeders have given us hundreds of new varieties that have been bred for greater pest resistance, hardiness, and fruit size. And only a few bushes will provide a family with all the fruit they will need for eating fresh or preserving. The plants fit easily into the home landscape in cluster plantings, as hedges and screens, or as single specimen plants. In addition to the beautiful sky blue berries in summer you can enjoy white cascades of lily-of-the-valley-like flowers in spring and radiant crimson foliage in the autumn. The blueberry is truly a plant for all seasons.

CLASSIFICATION AND ORIGIN

The fruit of many species of wild blueberry are harvested for local use, but those of three species—*Vaccinium corymbosum* (highbush blueberry), *Vaccinium ashei* (rabbiteye blueberry), and *Vaccinium angustifolium* (lowbush blueberry), are today sold on a large commercial scale. We'll deal primarily with the highbush and rabbiteye species in this chapter, since the lowbush is difficult to establish and not as easily cared for in the home planting.

Because wild berries were abundant, no serious effort was made to domesticate the highbush blueberry until the early twentieth century. In 1906, Dr. Frederick V. Coville of the United States Department of Agriculture initiated extensive research into the cultural practices best suited to this plant. His efforts, and those of his colleagues, led to the release of dozens of improved selections, many of which became the parents of varieties grown today. Tens of thousands of acres are grown in commercial plantings now, and the blueberry is rapidly becoming one of the world's most popular fruit. And well it should. Although only about eighty years from the wild state, today's varieties produce larger, juicier, and sweeter fruit on bushes that are more pest tolerant and better adapted to site variations than those of just a few years ago.

The rabbiteye blueberry, a southern species, has been cultivated even longer than the highbush blueberry. Important work by southern plant breeders has given us many new and improved varieties of this species, including a new group of low chill plants called "southern highbush" blueberries.

HUMAN NUTRITION

Blueberries have a wonderful taste and are quite nutritious. They are low in calories and sodium and are a good source of fiber and pectin, both known to lower blood cholesterol levels.

A cup of fresh berries weights about 145 grams, or five ounces, and contains twenty one grams of carbohydrate, one gram of protein and a half gram of fat, nineteen mg of Vitamin C, 145 International Units of Vitamin A, and only eighty five calories! The fruit also contains measurable quantities of ellagic acid which is known for its cancer-fighting abilities. And the juice of the blueberry and its relative, the cranberry, helps our bodies to ward off infections.

All in all, blueberries are a valuable and healthy part of our diets.

PLANT CHARACTERISTICS

The highbush blueberry plant is made up of several woody canes arising from buds in the crown to form a bush six to eight feet high. Rabbiteye plants can form bushes up to fifteen feet in height. The dense, fibrous root system is relatively small, reaching depths of only about three feet, depending upon soil conditions, though most of the important feeder roots are located in the top ten inches of soil. Most of the root system spreads only about as far as the dripline of the bush. Flower buds are set in the late summer and fall only on new shoots and bloom the following spring. Older wood bears no flower buds and is used only to bear leaves to support plant growth and to conduct water and nutrients to the top of the plant and carbohydrates to the roots.

ADAPTATION

The highbush blueberry can be split into two subgroups—the northern highbush and the southern highbush. The former is best suited for growing in northern areas of the U.S. where soils are acid. Large areas east of the Mississippi river, from North Carolina north to New England, and in the Pacific Northwest, are well suited to their culture. The southern highbush blueberry, developed from hybridization of the northern highbush with southern species, is well adapted to planting in South Carolina, Georgia, Florida, the Gulf States, and eastern Texas.

Rabbiteye plants, also members of a southern species, will produce crops from Virginia south along the Atlantic and Gulf coasts to Texas and inland to southern Arkansas, midland Tennessee and mountainous northern Georgia. These plants are more tolerant of upland mineral soils than the highbush, and considerably less soil modification is needed to grow them.

Late ripening varieties of highbush need a growing season of about 160 days, though early and mid-season varieties may need only 120-140 days. Flower buds can be damaged by temperatures of − 15°F. Though the well-hardened wood can tolerate slightly lower temperatures, don't plant the northern highbush blueberry in any location where winter temperatures are likely to fall below − 20°F, unless there is usually at least six inches of snowcover to insulate plant crowns during the coldest parts of the winter. For extreme northern areas consider the newer "half-highs." These are hybrids of the highbush and the lowbush blueberries that produce small plants bearing very high quality fruit. Because of their small stature they are easily covered with snows and protected against extreme cold. The degree of cold hardiness depends upon the species and variety of blueberry (Table 1).

Very hot areas of the country, with many days of temperatures above 90°F, will not support good northern highbush blueberry growth because the plant's root system cannot absorb water fast enough to replace that lost in transpiration. In other words, the plant dries out.

For warmer areas, consider the southern highbush or the rabbiteye blueberry, both of which are more tolerant of heat and dry soil conditions. Neither will tolerate extreme cold, and dormant plants of rabbiteye can be killed at temperatures of only − 10°F, again depending upon variety.

Avoid planting sites very close to buildings or surrounded by

TABLE 1. Some common varieties of blueberry listed in approximate order of decreasing cold hardiness.[z]

Highbush		
Hardy	Medium Hardy	Tender
Patriot	Jersey	Collins
Northland	Burlington	Berkeley
Meader	Rubel	Coville
Bluecrop	Earliblue	Pemberton
Blueray	Rancocas	Dixi
Elliott	Weymouth	Stanley
Herbert	Atlantic	Corcord
Bluetta		Southern Highbush
Northcountry		
Northsky		
Northblue		

- - - - - - - - - - - -

Rabbiteye		
Briteblue		Delite
Tifblue		Woodward
Climax		Homebell

[z] Most of this ranking is adapted from publications of the United States Department of Agriculture. Other researchers may rank varieties in a slightly different order.

hills or dense stands of trees. These have relatively poor air circulation and can increase plant damage from frost and diseases. However, they may also reduce winter damage by sheltering plants from strong winds. You decide which of these factors are most important in your situation.

Most blueberry flower buds of the northern highbush require about 900 hours (about two months) of chilling below 45°F, and will normally complete their rest requirements by mid-winter. Buds of the southern highbush and the rabbiteye require less than half that number of hours. Warm spells after the chilling requirement has been satisfied, with temperatures above 28°F, can damage the bushes by causing some growth activity in the flower buds and wood, making them more susceptible to subsequent cold temperatures. A rapid temperature drop following a midwinter warm period can damage buds and young wood. Because the conducting tissue is injured, shoots may appear normal, leaf out, and set fruit, only to shrivel and die soon after.

If you have selected carefully the proper varieties for your location, and selected your site wisely, your plants should need little winter protection. A snow fence can be erected to protect the plants from strong drying winds and a little extra mulch will help protect the roots from severe cold.

Cool, wet springs can cause temporary leaf discoloration. New leaves on emerging shoots of most varieties will be slightly reddened, while new leaves on 'Earliblue' will turn orange-yellow. This discoloration disappears soon after temperatures warm.

VARIETY SELECTION

There are dozens of blueberry varieties, each with its own peculiarities and characteristics which make it more or less adapted to a particular region of the country.

Select varieties that ripen their fruit early, midseason, and late to increase the length of the harvest season. Each variety will ripen its fruit over a several week period. And although blueberries are mostly self-fertile you'll have more fruit of the highest quality by planting more than one variety and allowing for cross pollination. This is especially important for the rabbiteye blueberry.

Following is a partial list of promising varieties for various regions of North America (Table 2). Because of the large number of varieties available, and because of varying microclimates within each region of the country, consult with your Cooperative Extension Agent and ask local growers their suggestions before making your final decisions.

Consider the following descriptions as a further aid in selecting suitable varieties. Each variety name is followed by an abbreviation to indicate whether it is northern highbush (N), southern highbush (S), or rabbiteye (R).

Avonblue. (S) The bush is small and spreading and must have

TABLE 2. Varieties of blueberries recommended for planting in different parts of North America.

Area 1: North Florida, Gulf Coast, eastern Texas, lower Southwest, California (south of San Diego).

Flordablue	Sharpblue	Avonblue	Woodard
Bluegem	Tifblue	Climax	Bluebelle

Area 2: Coastal Plain of Georgia, South Carolina (south of Charleston), Gulf states, east Texas, California (south of Los Angeles).

Flordablue	Sharpblue	Tifblue	Woodard
Southland	Delite	Briteblue	Climax
Bluebelle			

Area 3: Upper Piedmont and mountainous regions of Area 2.

Bluetta	Patriot	Berkeley	Blue Ridge
Croatan	Bluecrop	Murphy	Cape Fear
Morrow	Bluechip	Lateblue	Tifblue
Woodward	Southland	Delite	Briteblue
Climax	Bluebelle		

Area 4: Southern Virginia south to Piedmont and Coastal Plain Carolinas, Tennessee, lower Ohio Valley, southern and eastern Arkansas, lower Southwest, and mid-California.

Morrow	Croatan	Bluecrop	Southland
Murphy	Tifblue	Woodard	Climax
Bluebelle	Patriot (not for coastal plain Carolinas)		

Area 5: Mid-Atlantic states, Ozark highlands, Midwestern states, mountainous regions of Area 4, Washington, Oregon, northern California.

Bluetta	Lateblue	Collins	Darrow
Berkeley	Herbert	Bluecrop	Elliott
Blueray	Elizabeth	Patriot	Elizabeth

Area 6: New England and Great Lakes states

Earliblue	Berkeley	Bluetta	Darrow
Collins	Meader	Northland	Herbert
Patriot	Blueray	Bluecrop	Coville
Bluehaven	Spartan	Northblue	Northcountry
Northsky			

good soil. The high quality fruit ripen after 'Sharpblue' but before
the rabbiteye varieties. The plant is best adapted to north central
Florida conditions. Introduced–1977.

Berkeley. (N) The bush is upright, open-spreading, and easy to
prune and propagate. The clusters are loose and sometimes hidden
by heavy foliage. The berry is medium-large, light-blue, firm, mild-
flavored, slightly aromatic, and has medium dessert quality and a
large dry scar. It's liked for its beautiful color, firmness, and large
size but is inconsistently productive in some areas. Delayed harvest
may result in excessive fruit drop. The fruit ripens late-midseason.
Introduced–1949.

Bluebelle. (R) The plants are upright and moderately vigorous.
The light blue berries have a long ripening season, good flavor, and
a small scar. They begin ripening midseason, at about the same time
as those of 'Tifblue.' The berries often tear during harvest and will
therefore not store well.

Bluechip. (N) The bush is upright, easy to prune and resistant to
canker caused by *Botryosphaeria corticis*. The plant is also tolerant
to bud mites, mummy berry, and Phytophthora cinnamomi root rot.
The berry is very large, has excellent color, scar, and firmness, and
is pleasantly flavored. It ripens mid-June in Wilmington, N.C.
'Bluechip' has good ornamental value. Introduced–1982.

Bluecrop. (N) The bush is upright, drought-resistant, and consistently productive. The clusters are large and medium-loose. The
berry is large, very light-blue, slightly aromatic, and has medium
dessert quality. It has a small scar and resists cracking. The fruit
ripens midseason. This variety is the most widely grown in the world
and has excellent ornamental value to boot. Introduced–1952.

Bluehaven. (N) The bush is upright but may not be sufficiently
winter-hardy in the northern fringe of the blueberry production
areas. The berry is large, firm and light-blue with excellent flavor;
ripens early midseason, with 'Patriot' and 'Collins.' The scar is
very small and dry and berries hold their quality quite well on the
bush. It is susceptible to mummy berry. Introduced–1967.

Bluejay. (N) The bush is upright and slightly spreading. The
berry is medium-large, firm, light blue, mildly flavored and has a
small picking scar; ripens early midseason, with 'Collins.' 'Bluejay' has value as an ornamental. Introduced–1978.

Blueray. (N) The bush is very vigorous, upright, spreading, and consistently productive. The clusters are small and tight. The berry is very large, light-blue, firm, aromatic, of high dessert quality, with a medium scar, and is resistant to cracking. 'Blueray' has excellent ornamental value, often with pinkish or red-striped flowers. The fruit ripens midseason. Introduced–1955.

Blue Ridge. (S) The fruit of this variety are highly acid and medium to large. The variety is best adapted to Piedmont and the mountains of the southeastern US. Introduced–1987.

Bluetta. (N) The bush is short, difficult to prune, compact-spreading, moderately vigorous, and consistently productive. The berry is small to medium-sized, light-blue, firm, and has a broad scar, but softens rapidly if left too long on the bush. 'Bluetta' has good ornamental value. The fruit ripens early and has only fair flavor. Introduced–1968.

Brightwell. (R) The bush is upright and spreading and has a small crown. The berries are medium dark blue in color, medium in size and have a good flavor. They ripen over a relatively short period in about the same season as 'Woodard.'

Briteblue. (R) The bush is moderately vigorous and spreading. The berries are very firm, large, light blue, and very tart until fully ripe.

Climax. (R) The fruit of this plant resemble those of 'Brightwell,' but ripening begins a few days before the fruit of 'Woodard.' The ripening is concentrated and about 80% of the fruit can be harvested at one time.

Collins. (N) The bush is erect and fairly productive, though it sometimes is reluctant to send up replacement suckers. Fruit cluster is medium-sized and rather tight. The berry is medium to large, light-blue, firm, highly flavored, and does not drop or crack readily. 'Collins' has good ornamental value, with long lasting fall color. The fruit ripens early midseason. Introduced–1959.

Coville. (N) The bush is upright, spreading, and very productive. The cluster is loose. The berry is very large, medium-blue, firm, highly aromatic, attractive and tart when not fully ripe. It has very good dessert quality and scar and resists cracking. It does not drop readily and is excellent for processing. This variety is self-incompatible and domestic bees tend to prefer flowers of other varieties as well. The fruit ripens late. Introduced–1949.

Croatan. (N) The bush is upright and productive. The berries are large, dark blue, medium firm, slightly aromatic, sweet and have medium dessert quality. They ripen early, with 'Weymouth.' It is highly resistant to bud mite, moderately resistant to canker, and is the most widely planted variety in North Carolina. Introduced–1954.

Darrow. (N) The bush is vigorous and upright but not consistently productive in some areas. The cluster is medium-loose. The berry is large, light-blue, firm, aromatic, tart when not fully ripe, resistant to cracking, and has a good scar. The fruit ripens late midseason. Introduced–1965.

Delite. (R) The plants are moderately vigorous and upright. The fruit are medium large and often have a red undercolor visible through their heavy waxy bloom. Both ripe and underripe berries detach easily from the plant, so harvest must be delayed until most of the fruit are ripe.

Earliblue. (N) The bush is vigorous, upright, spreading, and fairly productive. The cluster is medium-sized and loose. The berry is large and aromatic, light-blue, firm, and has good dessert quality. It resists cracking, has a medium scar, and does not drop readily. It does not set sufficient fruit in some areas. It also may not be sufficiently winter hardy at the northern fringe of the blueberry production areas. It is very susceptible to phomopsis canker. The fruit ripens early. Introduced–1952.

Elizabeth. (N) The bush is upright and spreading. The cluster is loose. The berry is large and fair-colored with average flavor. The berries ripen over a long season, from mid-season to very late. Ripening begins just after 'Bluecrop.' This variety has not been productive in some northern areas. Introduced–1966.

Elliott. (N) The bush is winter-hardy and resistant to mummy berry. The berries are firm, light-blue, and medium-sized, with a good, mild flavor. The fruit may be very tart until 60% of the berries are ripe. They ripen very late, beginning seven to ten days after 'Lateblue.' Because of this, the variety may be too late for some northern areas. Cross-pollination is necessary for somewhat earlier ripening and larger size. 'Elliott' has excellent ornamental use. Introduced–1973.

Flordablue. (S) The bush is moderately vigorous and resistant to stem canker caused by Botryosphaeria corticis but very susceptible to Phytophthora root rot. Be sure to plant only in well-drained soil. The fruit are large, light blue, medium firm, and have a medium scar and good quality. The variety is best adapted to central and south central Florida. Introduced–1976.

Herbert. (N) The bush is vigorous and open-spreading, but inconsistently productive in colder climates. The cluster is medium-loose. The berry is very large, medium-blue, aromatic, tender, and has excellent dessert quality. It has a medium scar, and resists cracking. The fruit ripens late midseason. Introduced–1952.

Lateblue. (N) The bush is erect, vigorous and consistently productive. The berry is medium-large, firm, light-blue, aromatic, tart when not fully ripe, and has a small scar. 'Lateblue' has good ornamental value. The fruit ripen late. Introduced–1967.

Meader. (N) The bush is upright, open-spreading, winter-hardy, and consistently productive. The clusters are loose. The berries are medium-large, firm, and resist cracking, are uniform in size, of medium-blue color, and average flavor. The scar is small and dry. It ripens early to mid-season, with 80% of harvest during the first picking. It may tend to overbear. 'Meader' is particularly good where winter temperatures drop below $-20°F$ and heavy snow is common. The fruit ripen midseason. Introduced–1971.

Morrow. (N) The bush is slow growing, semi-upright and productive. The berries are large, light blue, and of high dessert quality. They ripen early and the season is short. Introduced–1964.

Murphy. (N) The bush is vigorous, spreading and productive and has some resistance to canker. The berries are large, dark blue and have fair dessert quality. They ripen early. Introduced–1950.

Northblue. (N) The bush is very low (2 feet) and blooms a week later than 'Bluecrop,' though the high quality fruit ripen a week earlier. Recommended for extreme northern areas.

Northcountry. (N) The bush is less than forty inches high and bears very light blue, medium sized fruit ripening about 10 July at South Haven, Michigan. It must have good cross-pollination. Recommended for extreme northern areas.

Northland. (N) The bush is four feet high and moderately spreading. The canes are flexible and tend to bend, but not break, under a snow load–an important feature in northern areas. It is one of the most consistently productive, ripening very early, with 'Earliblue.' The berry is small to medium-sized, firm, medium-blue, and has good flavor. The scar is small and dry. 'Northland' has very good ornamental use. Introduced–1967.

Northsky. (N) The bush is very low (10 to 18 inches in height) and bears medium sized, light blue fruit ripening about 15 July at South Haven, Michigan. It must have good cross-pollination. Recommended for extreme northern areas.

O'Neil. (S) The bush is vigorous and semi-upright. The fruit are very large, medium blue, firm, and have excellent, aromatic flavor. Flowering occurs early and continues over an extended period. The fruit ripen very early. This variety is a good choice for planting in coastal and piedmont regions of North Carolina. Introduced–1987.

Patriot. (N) The bush is four feet high, upright, open-spreading, and vigorous. The berries are large and firm, have good color and very good flavor when completely ripe. The scar is very small and dry. This is a good variety for northern areas because of its resistance to root rot and its high degree of winter-hardiness. It blooms early and is therefore susceptible to late frosts. The fruit ripen early to mid-season, usually with 'Collins.' Introduced–1977.

Sharpblue. (S) The bush is similar to that of 'Flordablue,' fast growing and vigorous. It has the lowest chilling requirement of any highbush and is very popular in Florida. The fruit are dark blue and have a small scar. The skin sometimes tears at harvest. Introduced– 1976.

Southland. (R) The plant is moderately vigorous and compact, producing many suckers. The fruit is light blue and firm, but the skin may become tough late in the season.

Spartan. (N) The bush is vigorous and partially resistant to mummy berry. The berry is medium to large in size, light blue, and highly flavored; ripens early midseason, with 'Patriot' and 'Collins.' Fruit size decreases rapidly after the first two harvests. Frost may be a particular problem with this variety and the plant is very sensitive to poor soil drainage and in some cases may do poorly on amended mineral soils. 'Spartan' has good ornamental use. Introduced–1977.

Tifblue. (R) The bush is vigorous and upright. It has a slightly longer chilling requirement than other rabbiteye varieties and doesn't do well near Gainesville, Florida. The fruit are very light blue and very firm. Don't harvest them until they are fully ripe. This is the hardiest of the rabbiteye varieties and does especially well when pollinated with 'Woodward.'

Woodard. (R) The plant is moderately vigorous and suckers freely. The fruit are medium firm and light blue and have a strong aromatic flavor. They ripen about a week before those of 'Tifblue.'

SITE AND SOIL REQUIREMENTS

Blueberries grow best in full sun on moist, acid soil. Wild huckleberry, laurel, blueberry, or a mixture of pine, red maple, and white cedar growing near your planting site generally indicate the right soil conditions for blueberries. Soil pH can range from 4.5 and 5.2 but the optimum pH is about 4.8. Rabbiteye blueberries are not quite so fussy in their soil requirements, though they still prefer an acid soil. Soil pH higher than about 5.5 can result in an iron deficiency, causing the leaves to yellow, and can convert soil nitrogen from the easily available ammonium form to the less available nitrate form. Soil pH lower than about 4.0 might allow aluminum and manganese to become available to the plant in toxic amounts, resulting in poor bud break and some dieback.

Adjust the soil pH before planting to fall to within the acceptable range. Use one of the soil acidifiers recommended in chapter one, but *do not use aluminum sulfate.* This adds no nutrients to the soil and cause result in a buildup of toxic aluminum. Fertilizing with ammonium sulfate will also acidify the soil over time, as well as add valuable nitrogen. Peat moss worked into the soil will make it more acid over a period of years.

The delicate, hair-like blueberry roots cannot penetrate a heavy, compact soil. Therefore, a soil that is moist, porous, well-aerated, and high in organic matter is best. Never plant blueberries in extremely wet soil where the water table is less than twenty inches below the soil's surface. In addition to restricting root growth, many wet soils, particularly in the south, are infested with the fungus *Phytophthora,* which causes root rot. If you have only a wet area for planting, position the plants upon small mounds of good soil to elevate their roots above the water line, or plant them in raised beds. For the blueberry plant to grow well, a hole twenty inches deep, filled with water, should drain within one hour. Most of the rabbiteye blueberries are tolerant of the root rot fungus but still don't do well in areas with poor drainage.

CULTURE

Soil Preparation

Prepare the soil at least one season in advance, as described in the introductory chapter, to thoroughly incorporate fertilizer and organic materials and subdue weeds. Pay particular attention to building up the organic matter. However, don't use maple or elm leaves if soil pH is above 5.0; they leave an alkaline residue upon decomposition.

Rotted sawdust or peatmoss uniformly mixed 50/50 (V/V) with good loam and used to fill the planting hole will increase plant survival, vigor, and fruit production. Soak the sawdust and peatmoss before incorporating or it will act like a sponge and dry the root zone. The organic material MUST be mixed with the soil; filling the hole with plain peatmoss or sawdust can damage plants.

Propagation

You'll always be better off purchasing certified disease-free plants from a nursery. It's easier than propagating and you can be sure of starting out with the best plants. However, some gardeners have a lot of fun propagating their own plants.

The propagating medium should be an equal mixture, by volume, of peat and vermiculite or coarse builder's sand. Never use beach sand. Place the medium in the tray or bed about a week ahead of time and wet it thoroughly. Dry peat is very difficult to wet; you may have to use hot water.

Propagating beds for hardwood and softwood cuttings may be made in any convenient size with six to eight-inch high sides (Figure 1). Construct miniature beds for use indoors using arched coat hangers and polyethylene plastic sandwich wrap, but remember to keep the beds out of direct sunlight (Figure 2). Cuttings will root best under cool white fluorescent lights on sixteen hour cycles.

Highbush blueberries are most commonly propagated by hardwood cuttings, but semi-hardwood and softwood cuttings, mounding, layering, leaf-bud cuttings, clump division, and budding also work. Rabbiteye blueberries do not root well from hardwood cuttings and are most often propagated by softwood cuttings and sucker division.

The propagating procedure is not difficult, but requires strict attention to details. Always take your cutting wood from healthy plants. But beware, however, that just because a plant looks healthy doesn't necessarily mean that it is healthy. Whenever you propagate your own material you are taking a risk of getting diseased plants.

FIGURE 1. A propagating frame. The hinged, screen covers can be lowered on bright, sunny days.

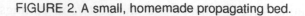

FIGURE 2. A small, homemade propagating bed.

Hardwood Cuttings

Hardwood cuttings are made from dormant, healthy, well-matured pencil-thick shoots of the previous season's growth. They may be taken after leaf drop in the fall, packed in damp sphagnum moss, sawdust or similar material, and stored in the refrigerator. Wood also may be taken in the spring before bud growth begins.

Cuttings are four to five inches long, pencil-thick, and taken near the middle or the base of the shoot. Make bottom cuts flush just below a bud, and the top cut at a slant and about one-quarter inch above a bud. Place the cuttings vertically in the rooting medium two inches apart, with two inches between rows and only the topmost bud showing. Treatment of hardwood cuttings with a root-promoting chemical generally has not been successful.

After setting the cuttings, thoroughly water to settle the medium around them. Leaves on the cuttings will develop quite early, and maintenance of high humidity is essential until the roots form. The roots do not form until after the first flush of top growth is complete, which may take two to three months.

When the second flush of top growth begins, roots have formed and the cuttings should be fertilized. Use a liquid fertilizer, commonly a 16-32-16 or 20-20-20 analysis, at the rate of about two level tablespoons per gallon of water. One gallon should be enough for about twenty five square feet of surface. Rinse all foliage with water immediately after application. Make weekly applications until mid-August.

You can leave rooted cuttings in the propagating bed all winter or transplant them to the nursery bed in early fall.

In the nursery bed, rooted cuttings are set about one foot apart in rows spaced about eighteen inches apart, and mulched. If soil is fertile you'll need no fertilizer the first year. After one year in the nursery, plants are considered two years old.

Semihardwood Cuttings

If you prefer to propagate with semi-hardwood cuttings, take them in the early summer while the plant is actively growing. The best time is just before the second flush of growth starts.

Cuttings should be four inches long with all the leaves but the upper two removed to reduce transpiration. The lower end of the cutting is scraped, moistened, and dipped into a rooting compound, such as a talc formulation of 4.5% indolebutyric acid, to promote rooting, then inserted into a rooting medium down to the lowest leaf. Care for these cuttings as you would hardwood cuttings.

SELECTION OF PLANTS

You can purchase blueberries as rooted cuttings or as plants one to five years old, though it's best to purchase vigorous, disease-free, dormant two-year-old plants twelve to eighteen inches high.

Buy your plants from a reputable nursery to assure trueness to name and freedom from disease. Nurseries specializing in blueberries offer a wide selection of varieties at reasonable prices. Order your plants at least six months before planting to assure the greatest probability of getting the varieties you want. A point to remember is that varieties that ripen their fruit later usually bloom later also. Consider these if you must plant in an area prone to late frosts.

Plants are sold as balled-and-burlapped (B&B), canned (potted), or bare-rooted. The first two can be planted with the least disturbance of the root system. Bare-rooted plants are usually shipped in peat moss wrapped in plastic. Open the package and remove the plastic as soon as plants arrive.

Planting

Blueberries can produce a satisfactory crop in locations shaded for three to four hours a day, but they'll produce their best crops when planted in full sunlight.

Even though fall plantings may result in more rapid spring growth, gardeners in most northern areas should plant in the early spring as soon as the ground can be worked. If you cannot set plants immediately upon delivery, unpack them and heel them in by placing the roots in a hole or trench and mounding soil around the entire plant, leaving only the top few buds exposed. Soak the roots for several hours prior to heeling-in if they appear dry. If the ground is frozen, put the plants in a cool, well-protected area, such as an unheated basement or a garage, and cover them entirely with damp peatmoss or sawdust.

Planting on a cloudy afternoon is best but not always possible. Set highbush plants at the same depth as they were in the nursery, about five feet apart, in rows spaced about ten feet apart. Because they are larger when mature, set rabbiteye plants at the same depth they were in the nursery, but about six to eight feet apart in rows spaced twelve to fourteen feet apart. Dig planting holes deep enough and wide enough so that no root crowding will occur. Usually, no root pruning is required at planting. If some larger roots on older plants are broken, however, cut them above the break. Trim excessively long roots—never wrap them around the root ball to fit into the hole. If you bought containerized plants, break up the rootball gently before planting. If plants are balled and burlapped (B&B) with real, old-fashioned burlap, untie the top of the burlap after planting. It is not necessary to remove the burlap completely. Plants wrapped in the newer plastic weave materials that resemble burlap should be removed from their wrapping prior to planting.

After setting the plant, fill the hole three-quarters full with soil or the soil mix and flood it. When the water has soaked in, fill the rest of the hole with soil, tamp firmly and water again.

Remove all flower buds at planting. Allowing the plant to bloom in the first year will interfere with its vegetative growth and establishment.

Blueberries are poor competitors for water and minerals. Allow no other plants to grow within two or three feet of the base of each plant for at least the first two or three years. A clean organic mulch around the base of the bush will discourage the growth of weeds, conserve moisture, and moderate soil temperatures.

POLLINATION

The blueberry flower resembles an inverted urn (Figure 3). Because of genetic, climatic, and cultural factors, a blueberry planting of many varieties can be in bloom for as much as a month. They are pollinated by bumblebees and honeybees. Bumblebees are the better pollinators because they fly in more inclement weather than honeybees.

Most varieties are self-fruitful and a planting will usually produce satisfactory crops when only one variety is present. Cross-pollination by several varieties, however, will give you more, larger, and earlier ripening fruit. This is especially important with the rabbiteye blueberry. Plant different varieties no more than twenty feet apart. If pollination doesn't occur within about three days after the flower opens, fruit set is unlikely. In general, there is adequate pollination activity if you observe fifteen to twenty bee entries into blueberry flowers within a ten minute period. Pollinated flowers remain white and drop within four to five days after bloom; flowers that have not been pollinated within the critical few days after they open remain on the bush for one to two weeks and turn a brilliant wine color. The flowers of 'Earliblue,' 'Coville,' 'Stanley,' 'Dixi,' '1316-A,' 'Berkeley,' and 'Jersey' are relatively unattractive to bees. Be sure to interplant these varieties with others for cross-pollination.

FIGURE 3. A blueberry flower looks like an inverted urn. The ovary develops into the berry after pollination and fertilization.

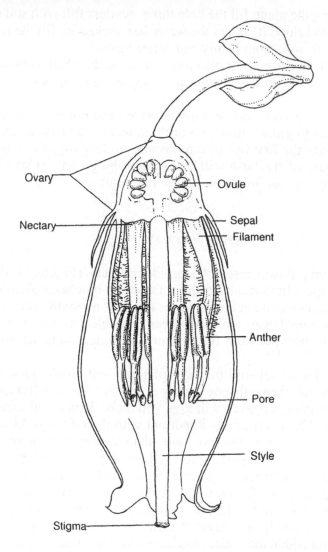

SOIL MANAGEMENT

If the planting is not mulched, shallow cultivation (less than two inches deep) throughout the season will help aerate the soil and control weed growth.

Mulching is the wisest and most widely used soil management practice in the home fruit garden. Maintain at least six inches of organic mulch. Grass clippings, peat moss, buckwheat hulls, shredded leaves, straw, wood chips, and rotted sawdust make good mulches. Don't use Grass clippings from areas previously treated with herbicides. Also, to avoid fermentation, overheating, and possible injury to the plant, don't use fresh grass clippings. Peat moss and buckwheat hulls are relatively expensive mulches. The former may crust on the surface, reducing moisture penetration to the plant roots, and the latter may blow away in a strong wind. Any organic mulch can increase the chances of mice injuring the crown of the plant in the winter. Of the last three materials mentioned, sawdust and wood chips are the best.

Under sawdust and chips, soil acidity remains nearly constant, the ammonium form of nitrogen used best by the blueberry increases, soil temperatures are modified, soil moisture is higher, and crop yields are usually substantially greater.

Spread sawdust and chips over the whole planting, or at least in a two-foot-radius circle beneath individual bushes, and slant it toward the plant to decrease wasteful runoff. Plan to use about five bushels of sawdust or chips per mature plant. This is equivalent to about ten, five gallon buckets. Mulching 1,000 square feet to a depth of six inches requires about nineteen cubic yards of material.

To reduce root desiccation, be sure the soil is thoroughly moistened before applying the sawdust. Otherwise, dry sawdust will soak up available soil moisture like a sponge and possibly dry the roots. Because soil microorganisms under organic mulching conditions use some nitrogen to break down the mulch, a nitrogen deficiency can develop. To prevent this, apply one pound of ammonium nitrate, or equivalent, for each 100 pounds of mulch used.

Fertilization

Apply fertilizer in a broad band completely around the plant beneath the dripline. Proper placement is important since it does not move horizontally in the soil to any appreciable extent. Raking in the fertilizer will help it to become quickly available for plant use. Further, there is little lateral transport of water and nutrients from one side of the bush to the other, and fertilizer applied to one side of the plant will primarily fertilize that side only! Be sure to apply fertilizer completely around the bush. Browning of the leaf tips and margins may indicate too much fertilizer, placement of fertilizer too close to the plant crown, uneven (clumpy) distribution of the fertilizer, or application of fertilizer in dry weather. Effects of over-fertilization may be lessened by heavy watering, which reduces the concentration of fertilizer in the soil and encourages leaching.

Although fertilizer recommendations vary according to soil and location, most experts agree on 1-1-1 ratio mixed fertilizer, such as 10-10-10. Avoid compounds such as nitrate of soda, calcium nitrate, bone meal, and wood ashes because of alkaline residues. Most other fertilizer compounds, such as ammonium sulfate and complete fertilizers, leave acid residues and are generally suitable for use on blueberry plants. Rabbiteye bushes respond well to applications of azalea-camellia fertilizer or cottonseed meal. Do not use fertilizers containing muriate of potash (KCl) since under some conditions, the chlorine may prove injurious to blueberry plant growth.

At least half the nitrogen supplied by the fertilizer should be in organic form, which becomes available to the plants over a long period of time.

Well-rotted barnyard manure (six inches deep) or poultry manure (one to two inches deep) may also be used, but should be applied only in late fall or very early spring, and never when the ground is frozen.

Fertilizing Young Plants

After the highbush plant has started a second flush of growth, often after two to three weeks in the field, fertilize it with the equivalent of about one ounce (two tablespoons) of a 10-10-10 fertilizer per bush. Spread this around the plant at least six inches, but not more than twelve inches from the crown. Make a second application in early July in areas with moderate winters.

Young rabbiteye plants are very sensitive to fertilizers. If you've prepared the soil properly and incorporated plenty of organic matter, you should not need to fertilize them the first year.

Mature Plants

Increase fertilizer applications each year until mature bushes (after six years in the ground) are receiving about one pound per plant, one-half applied at the beginning of bloom, the other five to six weeks later.

Because growing conditions throughout the country vary, you must establish your own rate of fertilization by watching for signs of good plant vigor and for certain symptoms of nutrient deficiency. In general, mature plants should produce several whips near ground level and laterals between twelve and eighteen inches long. Most productive shoots have between fifteen and twenty leaves. Poor vigor and leaf discoloration generally indicate insufficient fertilization. The following list of nutrient deficiency symptoms most often found in blueberry plantings can serve as a guide to possible plant problems.

Nitrogen Deficiency

Uniform paleness or yellowing develops over the entire leaf surface, followed by reddening and death of the leaves. Older leaves show symptoms first. Young shoots arising from the base of the plant appear pink, then turn pale green. The entire plant is severely stunted.

Treatment: Apply a nitrogen fertilizer, preferably with nitrogen in the ammonium form.

Iron Deficiency

A lemon-yellow color develops in the area between the leaf veins (interveinal chlorosis). Veins remain green. In severe cases, the leaves may turn completely yellow, or sometimes reddish-brown. Basal leaves will be stunted and new shoots yellow in color. Terminal leaves will generally discolor first.

Rabbiteye plants are more susceptible to iron deficiency than highbush plants.

Temporary Treatment: Foliar spray of ferrous sulfate (34% Fe) or soil treatment with iron chelate (one and one half ounces for new bushes, four ounces for established bushes).

Long-Term Treatment: Adjust soil to the proper pH and apply fertilizer with the bulk of nitrogen in the ammonium form.

Magnesium Deficiency

Symptoms usually appear at berry-ripening time. There is pale green coloring between the veins and on margins of lower leaves of very vigorous shoots. In severe cases, the affected leaves turn yellow to olive-green with interveinal areas orange to red. Treatment: Apply magnesium sulfate (Epsom salts) or sulfate of potash-magnesium to the soil at the rate of about five pounds per 1000 square feet. Applications may have to be repeated every three years.

In addition to visual inspection of foliage for nutrient deficiency symptoms, periodic soil sampling and foliar analysis are useful for determining the health and vigor of the plants.

When you have your soil tested, ask that magnesium and calcium tests be made also. A magnesium to calcium ratio of 1:10 and a potassium to calcium ratio of 1:5 are usually considered about right for blueberry soil.

What appear to be symptoms of mineral deficiency might be the result of an actual deficiency or the result of several other conditions or combinations thereof. These may be:

1. Insufficient or poor soil moisture distribution, topsoil erosion.
2. Poor drainage and subsequent restriction of the root system.
3. Insects, disease, fertilizer burn, weeds, or compaction of the soil–all of which can weaken the root system.
4. Insufficient quantities of available ammonium nitrogen.
5. Periods of cool weather during the growing season.
6. Injury from pesticides or weed control materials.

Be sure that some of the above conditions don't exist in your planting before applying additional fertilizer.

Watering

Blueberries require a constant supply of moisture. A mature planting should receive about one to two inches of water per week during the growing season, or at least from fruit set through harvest. Gardeners often employ the "feel test" to determine the moisture needs of the soil. Squeeze a sample of soil in the palm of your hand. If soil moisture is adequate, the ball of soil thus formed is weak and easily broken. If the soil ball is not easily broken, the soil may be too wet; if the soil will not form a ball, it is too dry.

Typical drought symptoms include reddened foliage, weak, thin shoots, early defoliation, and decreased fruit set. Don't use overhead sprinklers during ripening because steady wetting can cause ripe fruit to split. Early morning sprinkling with good drying weather, however, is quite safe. You can also use this type of sprinkling for frost protection during bloom. In that case, begin when the temperature drops to just above freezing and continue sprinkling until the temperature rises and the ice melts. This practice can protect the blossoms down to 25°F.

Trickle irrigation with a soaker hose is probably the best way to water plants because it reduces the amount of water lost through wasteful evaporation and supplies water directly to the soil where roots can take it up. It also uses less water than sprinklers and avoids potential damage to the fruit through wetting.

Fortunately, if you mulch the planting with a good organic mulch you may not need to water more than a couple of times during the growing season.

PRUNING

You must know the blueberry's growth and bearing habits in order to prune it correctly.

The plant begins vegetative growth with bud swell in the early spring and continues into late summer or early fall. Shoots grow in a "flushing" manner, that is, a shoot will make rapid growth, then stop when the apex dies. After a short period of time, a bud near the top of the shoot will break and continue the shoot growth. Each shoot may have a single, or several, growth flushes during the season. Finally, in mid to late summer, shoot growth ceases, and buds that previously would have broken into shoots change, or differentiate, into flower buds (Figure 4). Each flower bud contains from five to ten flowers. The buds begin to form at the tip of the shoot. By this time, the plant has begun to enter its rest period, which usually inhibits the flower buds from blooming immediately. This "rest" is normally broken only after the plant has been subjected to several hundred hours of temperatures below 45°F. Warm spring weather reactivates growth and development of the buds

Start pruning immediately after setting your new plants into the ground and continue it each year in the early spring before the buds begin to swell. Some gardeners prefer to prune in the fall because weak canes can be easily identified by the fact that their leaves turn color first. The time of pruning affects the blooming period. Plants pruned in the fall bloom later than those pruned in the winter or early spring, and this will decrease the chances of spring frost damage. However, fall pruned plants may also suffer greater winter injury. Plants not pruned at all will bloom the earliest and are therefore most susceptible to late frost damage. Never prune during budswell or bloom—swollen buds are easily broken off and pruning during bloom greatly retards the vegetative growth of the bush.

The severity of the pruning can influence ripening. Light pruning prolongs the ripening season, while heavy pruning concentrates it. In addition, very heavy pruning will substantially decrease the yield (in pounds) of fruit per plant. The size of individual fruit, however, is increased.

Though rabbiteye plants may need pruning only once every several years, highbush plants are usually pruned annually. Because blueberry forms its flower buds on current year's wood and blooms on one-year-old wood only, each year's crop is borne farther from the root zone and the bush center. This means that each year nutrients must travel a longer distance to fruiting wood as well as supply much non-fruiting, "extra" wood. Some of this non-fruiting wood does support the leaves and is necessary for best growth. Excessive amounts of it, however, can "drain" some nutrients that would otherwise benefit fruit development. The purpose of pruning is to remove excessive amounts of this unproductive wood. This will stimulate the growth of vigorous new shoots that will bear the

FIGURE 4. Dormant blueberry shoot. Note the larger flower buds near the top and the smaller, pointed vegetative buds below.

following season's crops. Canes older than five years are not productive, nor are bushes with more than about eight to ten canes.
Therefore, remove all dead, injured, and weak canes, canes older
than five years, and crowding canes first.

In addition to reducing competition, pruning will also allow better light, air and spray penetration into the bush.

At planting. Remove all weak, spindly growth and flower buds
from newly set plants by tipping back the shoots. This will encourage healthy bushes, since early fruiting will delay the vegetative
growth necessary for large-sized bushes.

After one growing season. Allow the more vigorous bushes to
bear a small crop of less than one pint per bush (twenty to thirty
flower buds). Remove weak wood. If too many buds still remain,
thin out or head back the remaining bearing shoots. On smaller, less
vigorous plants, remove all weak growth and all flower buds (Figure 5).

After two seasons. Most bushes can bear a small crop of one to
two pints per bush. The emphasis, however, should still be on
establishing a healthy, vigorous bush and not on fruit production.
Heavy fruit production at this time will dwarf the plant. Again,
remove all weak and injured growth (Figure 6). Follow a similar
pruning procedure every year, allowing the plant to produce succes-
sively larger yields until the bush is mature.

If well-grown bushes were started as healthy two-year-old plants,
they can be considered mature bushes after six to eight growing
seasons in the field.

FIGURE 5. Virgorous 'Earliblue' blueberry bush one year after planting. Left,
before pruning; right, after pruning.

FIGURE 6. Vigorous 'Coville' bushes two years after planting. Left, before pruning; right, after pruning.

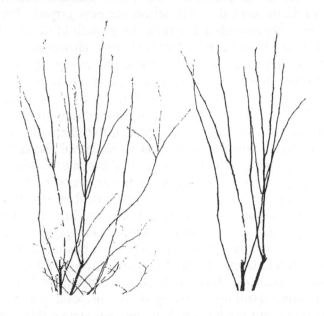

Mature plants. There are three points to consider in pruning a mature bush: (1) prune lightly enough to ensure a heavy crop for the current year, (2) prune severely enough to secure large-sized berries, and (3) prune enough to balance crop and bush vigor and thus assure sufficient new wood for future bearing and to contain overall bush size.

After removing all diseased, dead and injured wood, cut back old and unthrifty canes to ground level or to low vigorous side shoots. Don't cut back to stubs; these can harbor borers. Remove all soft, basal growth. Such growth is not stiff and springy but limber and often irregularly flat-sided instead of round. This type of growth is nonproductive and very susceptible to disease and winter injury. Next, remove all canes older than five years. These are usually about one and one-half inches in diameter at their base. You can keep ahead of the game by removing the oldest 20% of the canes each year.

Absence of new shoots frequently indicates that the bush is overcrowded with old canes and/or under fertilized. Removal of large (0.75-1.0 inches in diameter) and some medium-sized (0.5-0.75 inches in diameter) canes will stimulate new growth. Prune these bushes severely enough to promote the growth of three to five new canes from the base of each plant each year. However, take care not to stimulate too many new canes since this would necessitate additional pruning and would also result in excessive vegetation shading the leaves. The most productive bushes are those with eight to ten canes, with 75% of those canes being of medium thickness.

After the first two steps are completed, thin the bush by removing twiggy or bushy growth clusters and weak lateral shoots (Figure 7). Thin erect growing varieties more in the center, while spreading varieties require more pruning of the lower, drooping branches.

In pruning, always remove less vigorous, thin growth and leave thicker, more vigorous wood. Buds borne on thick wood open later in the spring and are less susceptible to late spring frost damage. They also produce larger berries. Unpruned bushes degenerate rapidly into a thick, twiggy mass of unfruitful wood (Figure 8 a,b).

In many instances, blueberry bushes are not pruned or are pruned insufficiently, resulting in overgrown, diseased plants with decreased vigor and production. You can rejuvenate these by cutting them back to the ground. The first summer after pruning there will

FIGURE 7. Left, the bushy growth on the left of the bush should be removed during pruning. Right, the short lateral side shoots of this bush will produce inferior berries and should be removed during pruning.

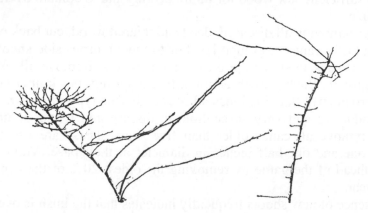

FIGURE 8a. Mature 'Earliblue' blueberry bush before pruning.

FIGURE 8b. Mature 'Earliblue' blueberry bush after pruning.

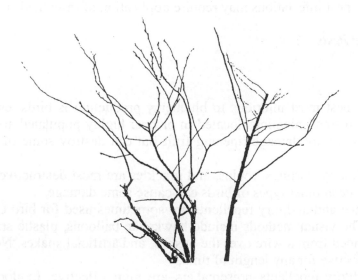

be no crop. However, the bushes should bear a substantial crop the following summer and should be in full production in the second or third year. An alternative is to cut back half of each bush and allow the other half to bear fruit. The following year, cut back the unpruned half. This procedure will completely rejuvenate each bush in two years while providing a partial crop each year.

PEST CONTROL

Generally, the homeowner is little plagued by pests in the home blueberry planting and many have cultivated blueberries for years without having to use any pest control measures. Still, some pests do attack these plants and controls must be used when infestations become damaging. Because of rapidly changing policies, any control recommendations may become outdated and illegal at any time. Always consult your local Cooperative Extension Service or state university for the latest pest control recommendations for your area.

In general, keep your planting clean and free from weeds and debris. If you see a single weakened plant that looks very sick, pull it out and destroy it. If you see caterpillars or other insects on your plants, pick them off and drop them into a can of water and oil. Severe pest infestations may require application of insecticides.

Larger Pests

Birds

The pest most injurious to blueberry production is birds, especially in small plantings located in or near highly populated areas. Birds not only damage ripening fruit but can destroy some of the buds.

Although starlings, robins, and grackles are most destructive, at least fifteen other types of birds can cause some damage.

Visual and auditory repellents are sometimes used for bird control. The visual methods include hawk-like balloons, plastic strips suspended from a wire over the bushes, and artificial snakes. None are effective for any length of time.

Auditory repellents—noisemakers—are more effective. Exploders of many kinds are on the market, but they are not as effective as electronic devices. They also annoy neighbors. Electronic devices transmit either a bird "distress call" or other sounds which repel birds.

The most effective method of bird control is to cover the plants with protective netting just before the berries begin to ripen. Remove it after the last harvest. A good-quality net, if used only during berry ripening and stored properly between seasons, can last ten years or more.

Rabbits and Deer

Both these animals sometimes browse the tips of stems in winter, destroying some fruiting wood. If the tips of low shoots are cut away clean, as though the cut were made with a knife, then rabbits are the culprits. If the cuts are ragged, as though the shoots were partially severed and then pulled the rest of the way off, then deer are feeding.

There are some fairly good rabbit repellents on the market that can be sprayed on the plants. Human hair is an excellent repellent for deer. Collect hair from the barber's and hang it in old nylon hose in the bush. Large plants may need a couple of these "hairbags" to keep the deer away.

HARVESTING

Blueberry plants can start producing in the second year after planting and continue in well-managed plantings for more than fifty years. Yields vary substantially depending upon soil conditions and management practices but will commonly average about one pint per bush in the third year, and may range up to about twenty pints per bush after eight to ten years. Some gardeners have reported yields as high as fifty pints per bush.

Fruit Ripening

The blueberry fruit requires two to three months to ripen. If berries are harvested as soon as they show a slight pink coloration, ripening will still continue, but the fruit will not be as sweet or as large as if it were ripened on the bush. If you allow the fruit to remain on the bush for an additional three to five days, they can accumulate up to 15% sugar and increase half again in size. A dead ripe berry will not have a pink ring around the scar, the point of attachment of the stem to the fruit.

Picking

Because overripe berries shrivel and drop, pick over the entire planting about once a week.

Blueberries do not ripen simultaneously in the cluster. Role ripe berries from the cluster into the palm of the hand with the thumb. This prevents immature berries from becoming detached and reduces tearing and bruising of the fruit. If you have a lot of plants you can rig a shoulder or belt fastener for the container so that both hands are free for picking. Excessive handling destroys the berries' attractive, whitish surface cover, or bloom, and increases bruising and spoilage.

USE IN THE LANDSCAPE

Attractive Qualities

Blueberries are among the most beautiful plants in our landscape. Their elegant white or pinkish blossoms in spring give way to bright blue fruit in summer. The plants join the maples in the autumn in flaming their leaves to a brilliant crimson or yellow. The stems themselves stand out red and yellow against the white winter snow.

Plants can be located in the landscape as specimen plants or put to use as hedges to divide areas of the property or as screens to afford privacy in suburban yards. Because of their relatively small stature, the plants are not obtrusive and will fit easily into nearly any area. The blueberry is truly a "plant for all seasons." Ornamental values of some of our more popular varieties are given in Table 3.

TABLE 3. Notes on the ornamental value of northern highbush and rabbiteye varieties.

Rating[z]	Variety	Notes
		Northern Highbush
G	Berkeley	Gracefully spreading
G	Bluechip	Stems amber-red in winter
E	Bluecrop	Fiery-red in autumn
VG	Bluehaven	Half high, upright
G	Bluejay	Yellow-orange in autumn
E	Blueray	Burgundy-red in autumn
G	Bluetta	Low stature
E	Burlington	Upright, bushy
G	Collins	Leaves redden early, last long
G	Coville	Yellow-orange in autumn
G	Earliblue	Attractive bush shape
VG	Elliott	Blue-green leaves turn orange-red
E	Friendship	Long lasting brilliant orange-red
G	Herbert	Stems yellow in winter
G	Jersey	Long lasting yellow-orange in autumn
VG	Northcountry	Fiery-red in autumn; outstanding
G	Northblue	Dark green leaves turn dark red
VG	Northland	Orange-red in autumn, yellow stems
G	Northsky	Bush 10-18 in.; lvs. dark red in autumn
E	NovemberGlow[y]	Slender leaves turn bright red
VG	Ornablue[x]	Profuse bloom, long lasting autumn color
G	Rancocas	Slender leaves red in autumn
VG	Rubel	Long slender leaves turn bright red
VG	Tophat[y]	Dwarf, profuse bloom, Bonsai type

TABLE 3 (continued)

Rating[z]	Variety	Notes
		Rabbiteye
VG	Beckyblue	Powder-blue foliage turns yellow-blue in autumn. Stems red in winter.
VG	Bluegem	Light green foliage turns red.
VG	Bonitablue	Bright red foliage in autumn.
G	Chaucer	Profuse bloom.
E	Choice	Outstanding autumn color and winter stem appearance.
G	Climax	Round leaves turn fiery red.
VG	Sunshine Blue[z]	Profuse bloom, very fine leaves.

[x] Rating: G = Good; VG = Very Good; E = Excellent.

[y] Strictly ornamental types. Not for fruit production.

[z] Southern highbush type.

Currants, Gooseberries, and Jostaberries

Danny L. Barney

Currants and gooseberries have been immensely popular fruits in northern Europe and Britain for centuries. From colonial days until the early 1900s, they were also popular in the United States and Canada, both commercially and in home gardens. When due care is given to site and variety selection, the plants are easy to grow and the fruits are suited to a multitude of uses. The introduction of white pine blister rust from Europe in the late 1800s lead to widespread restrictions on *Ribes* production in the United States, and so the fruits were lost to us for a generation. With the development of rust-resistant pines and *Ribes*, currants and gooseberries are once more becoming popular with home gardeners. Jostaberries were developed in the 1900s from crosses between black currants and gooseberries. Their disease resistance and vigor make them a promising crop for the future.

CLASSIFICATION AND ORIGIN

Woody shrubs in the genus *Ribes* are found throughout the northern portion of the northern hemisphere and along the Andes Moun-

Danny L. Barney is Associate Professor of Horticulture, University of Idaho and Superintendent, Sandpoint Research & Extension Center, 2105 N. Boyer, Sandpoint, ID 83864.

[Haworth co-indexing entry note]: "Currants, Gooseberries, and Jostaberries." Barney, Danny L. Co-published simultaneously in *Journal of Small Fruit & Viticulture* (Food Products Press, an imprint of The Haworth Press, Inc.) Vol. 4, No. 1/2, 1996, pp. 107-142; and: *Small Fruits in the Home Garden* (ed: Robert E. Gough, and E. Barclay Poling) Food Products Press, an imprint of The Haworth Press, Inc., 1996, pp. 107-142. Single or multiple copies of this article are available from The Haworth Document Delivery Service [1-800-342-9678, 9:00 a.m. - 5:00 p.m. (EST). E-mail address: getinfo@haworth.com].

tains in South America. There are about 150 *Ribes* species world-wide, 18 to 21 of which have been used in developing modern varieties. In addition to several species that are used as ornamentals in landscapes, five types of *Ribes* are grown for their fruit: black currants, red currants, white currants, gooseberries, and jostaberries.

Currants and gooseberries have been used for centuries as food and medicines, although they were apparently unknown to the Greeks and Romans. Currants were originally known as "corans" or "currans" in England, probably because their fruit resembled 'Corinth' grapes. Red currants were also known as "red gooseberries," and the English name "beyond-the-sea gooseberries," and the French and Dutch names "groseilles d'outre mer" and "overzee" indicate that the fruits may have been imported, possibly by the Danes and Normans. Regardless of where they originated, red currants were collected from the wild for use in medications as early as the 1400s A.D. Red and white currants were probably first cultivated as garden plants in Holland, Denmark, and the coastal plains surrounding the Baltic. During the 1700s, black currants were domesticated in eastern Europe and sold at farmers markets in Russia.

Currants were first introduced into North America in 1629 with the importation of the red currant variety 'Red Dutch.' In a memorandum dated March 16, 1629, the Massachusetts Company stated that it was providing for the interests of the colony in the New World: "To provide to send for New England, Vyne Planters, Stones of all sorts of fruites, as peaches, plums, filberts, cherries, pear, aple, quince kernells, pomegranats, also wheate, rye, barley, oates, woad, saffron, liquorice seed, and madder rootes, potatoes, hop rootes, currant plants."

White and black currants were probably introduced into North America as early as the reds. While white currants found some favor in the United States, black currants never became widely popular here. Early botanists generally included black currants in their discussions with red and white varieties, but were more often interested in the medicinal qualities of black currants, rather than their fruit quality. Because of their medicinal value, black currants were used in nearly all northern European countries. In England, black currants were once called "squinancy berries" because they were used to treat quinsy (tonsillitis). Today black currant juice and other products are very popular foods in Europe.

Gooseberries were probably first introduced into England during the reign of Queen Elizabeth (1533-1603). The credit for improving the gooseberry can largely be attributed to textile weavers in Lancashire, England who competed fiercely in producing prize-winning gooseberries. Charles Darwin reported that 171 gooseberry shows were held in England during 1845. Large berry size was the primary goal of the gooseberry fanciers, but flavor, beauty, and productiveness were also important.

Gooseberries were introduced into North America at about the same time as currants, but the European varieties were highly susceptible to powdery mildew, which created serious problems for growers. In 1833, the gooseberry picture changed with the discovery of a mildew-resistant seedling by Abel Houghton of Lynn, Massachusetts. This seedling was the first successful cross reported between European and North American gooseberries. The number of gooseberry varieties today is astounding. According to a 1969 estimate, there were 4,884 red, yellow, green, and white gooseberry varieties. Of those, only a few hundred are commercially available and worthy of cultivation.

The 1920 census showed approximately 7,400 acres of commercial currants and gooseberries in the United States. Most of the production was concentrated in the middle Atlantic, upper midwest, and northeast states, with small farms scattered throughout the rest of the continental United States. Nearly all of the production was made up of red and white currants.

Commercial production continued until the 1930s when blister rust became a major problem for the lumber industry in the United States. The disease organism, which had been accidently introduced from Europe in the late 1800s, can only survive when both susceptible *Ribes* and five needled pines are grown close to one another. Eastern white pine, western white pine, and sugar pine were valuable timber species threatened by the disease. The fungus must spend part of its life cycle on one host and part on the other. Black currants are the most susceptible *Ribes* host, while domesticated red and white currants and gooseberries are quite resistant to blister rust. Several *Ribes* species native to North America are susceptible to the disease, which lead to the establishment and continued existence of blister rust in the wild.

Because the disease threatened valuable forest resources and gooseberries and currants were minor crops, the federal government and many state governments took steps to prevent *Ribes* cultivation, and even to eradicate native currant and gooseberry species. Although the eradication efforts did reduce the incidence of the disease in some areas, most programs were considered unprofitable and eventually abandoned. In time, breeders succeeded in producing white pines and black currants which were resistant to blister rust. There are no longer any federal bans on growing currants or gooseberries, although 25 states continue restrictions on *Ribes* importation or cultivation today (Table 1).

European production continued despite the bans on currant and gooseberry production in the United States. Today, *Ribes* production, especially black currants, is a major industry in Europe. In North America, there are only about 200 acres of commercial gooseberries and currants, but there is interest in expanding production, and the fruits are receiving attention from home gardeners.

Jostaberries are hybrids between black currants and gooseberries, and were first developed in Germany in the 1940s. This is a new crop for which cultural practices are still being developed. Jostaberries appear to be resistant to white pine blister rust.

HUMAN NUTRITION

Gooseberries, currants, and jostaberries are rich in vitamins A, B, and C. In fact, black currants contain about 4 times as much vitamin C as citrus fruits. *Ribes* fruit also contain pectins, various mineral elements, and fructose. Currants are mostly used for making jellies, relishes, and juices. The leaves and buds of black currants are used for herbal medicines and the juice is often mixed with other fruit juices or used to make wine. Because of their high vitamin C content, the juices from red, white, and black currants are sometimes used as substitutes for citrus juices in parts of Europe. Except for sampling in the garden, currants generally aren't eaten out of hand because the berries are quite seedy. However, the tart flavor of fresh,

TABLE 1. States restricting *Ribes* importation and/or production.

The following states prohibit the importation or growing of currants and gooseberries. Contact your state department of agriculture at the number listed for more information.

NC	(919) 733-6930	NH	(603) 271-2561

The following states have some restrictions on importing or growing currants and gooseberries. Contact your state department of agriculture at the number listed for more information.

AL	(205) 242-2656	NJ	(609) 292-5440
AZ	(602) 542-0972	NY	(518) 457-7370
CO	(303) 239-4140	OH	(614) 866-6361
DE	(302) 739-4811	PA	(717) 787-4843
FL	(904) 372-3505	RI	(401) 277-2781
IA	(515) 281-5861	SC	(803) 656-3006
IL	(708) 990-8256	TN	(615) 360-0130
MA	(617) 727-3031	VA	(804) 786-3515
ME	(207) 289-3891	VT	(802) 828-2431
MI	(517) 373-1087	WA	(206) 586-5306
MT	(406) 444-3730	WV	(304) 558-2212
NE	(402) 471-2394		

Note: Many states have general regulations that control the importation of plant materials. Even if your state has no specific *Ribes* regulations, imported plants may have to be inspected and certified. Contact your state department of agriculture for details.

ripe red currants is very good. White currants are also good when eaten fresh, but are generally considered less flavorful than red currants. Because of their strong, "foxy" flavor, black currants generally aren't eaten fresh, but make excellent juices, syrups, and other products. One way to increase the appeal of black currants is to blanch them in boiling water for several minutes to remove some of the flavor and dark pigment. Rinse the berries and cook in fresh water to lighten the color and further subdue the flavor. My favorite black currant recipe is to make a thick, sweetened syrup for use on pancakes. Before thickening, strain the juice to remove seeds and pulp. Combining 3 parts of water and 1 part syrup makes an excellent drink.

Gooseberries are used primarily in preserves, pastries, compotes, and other desserts, but bush-ripened berries eaten out of hand are a delightful treat. Ripe gooseberries have a sweet, delicate taste and a texture resembling grapes. A few gooseberry varieties produce berries covered with fuzzy hairs that give fresh fruit an unpalatable texture, though most have smooth fruits.

Jostaberry fruits are smooth, sweet, flavorful, soft, and dark purple to nearly black. They are normally used for processing, but can be eaten out of hand. Like currants and gooseberries, jostaberries keep well when frozen.

PLANT CHARACTERISTICS

Jostaberries are the largest members of the genus *Ribes* cultivated for fruit. The plants can develop rank growth and may reach 8 feet (2.5 m) in height. Healthy, well-pruned currant plants are normally about 3 to 5 feet (90 to 150 cm) tall. American gooseberries range from 2 to 4 feet (60 to 120 cm) tall, with European types somewhat smaller.

The main branches of *Ribes* bushes are called canes, and arise from the crown (slightly below to slightly above the soil surface) each spring. Canes on cultivated currant and jostaberry bushes are smooth, while those on gooseberry plants have sharp spines at the nodes. New shoots developing from buds on the canes allow the canes to grow taller. Very short mature shoots are called spurs and often bear flower buds.

Red currants, white currants, gooseberries, and jostaberries set many of their flower buds on spurs located on two- to three-year old wood. Most flower buds on black currants are produced on one- and two-year old wood. Currant flowers and fruit are borne in long clusters called strigs, while gooseberry and jostaberry flowers and fruits are borne singly or in small clusters. Both gooseberries and currants bloom very early in the spring and are pollinated by bees and other insects. Figures 1 and 2 illustrate *Ribes* canes, flowers, leaves, and fruit.

Because they lack the taproot which develops in seedlings, nursery-propagated currants and gooseberries have shallower root sys-

FIGURE 1. *Ribes* canes and leaves (A) and (C) gooseberry bush and cane. (B) currant or jostaberry bush. (D) red or white currant cane. (E) black currant cane. (F) typical *Ribes* leaf.

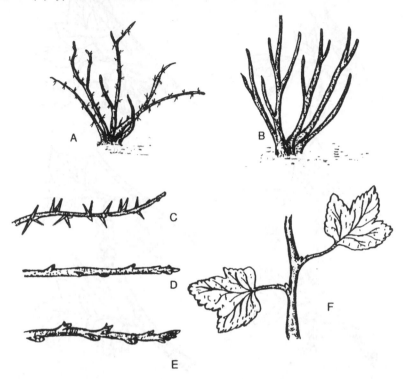

tems than seedlings, generally about 16 inches (40 cm) deep with a spread of 6 feet (2 m) or less. The root systems of tall, vigorous bushes may be larger. Feeder roots that absorb water and nutrients are especially abundant near the soil surface.

ADAPTATION

Domestic gooseberries and currants are native to cool, moist climates and most varieties require between 120 and 140 frost-free days. Although gooseberries generally tolerate high temperatures better than currants, the leaves and fruit can still be damaged by temperatures of 86°F (30°C) or higher. In order to flower and grow normally, currants, gooseberries, and jostaberries need to go through

FIGURE 2. *Ribes* flowers and fruits. (A) gooseberry. (B) red or white currant. (C) black currant.

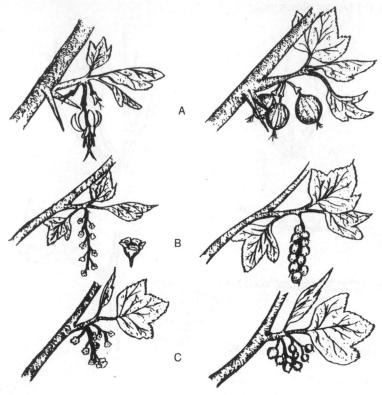

a dormant period during the winter. Dormancy is brought on by short days and cold temperatures, and a certain number of hours at temperatures near freezing are needed for *Ribes* to complete and break dormancy. Short, warm winters can prevent them from meeting their dormancy requirements, although European gooseberries may be somewhat more tolerant of these conditions than other cultivated gooseberries, currants, and jostaberries. The problems with heat damage and the difficulty in meeting dormancy requirements make *Ribes* cultivation difficult in southern climates. Figure 3 shows areas of the United States where currants and gooseberries are adapted. Before planting, however, make sure you can legally grow *Ribes* in your state.

FIGURE 3. Map of the United States showing suitable climates for currants and gooseberries. Domesticated currants and gooseberries are well-adapted to regions A and E. In regions B and C, the summers tend to be too long and hot. Low summer rainfall limits dryland production in region D, but *Ribes* can be grown with irrigation. Region F is too cold for reliable production, but the hardiest varieties can be grown in some areas. Note that these regions aren't sharply defined, but gradually blend into one another depending upon elevation and other factors.

While they aren't particularly tolerant of high temperatures, gooseberries and currants are very cold hardy (generally considered more cold hardy than apple trees) and some varieties, when fully acclimated, can tolerate winter temperatures of $-31°F$ ($-35°C$) or less. Unfortunately, there are many varieties available for which accurate cold hardiness data is lacking. In cold, windy areas, such as the Great Plains, windbreaks improve *Ribes* survival and performance. In the coldest parts of the country, apply sawdust or other organic mulch around the bushes to help protect the roots and crowns from freezing. This practice should ensure new cane production even if the old canes are occasionally killed.

Ribes are among the earliest-blooming fruits and are susceptible to spring frosts. The blossoms are killed or injured by temperatures around 26°F (-3°C) or less. Select a planting site with the best possible air drainage to reduce frost damage and disease problems. If your planting site experiences frost during the currant or gooseberry bloom, you can cover the plants or use irrigation sprinklers to reduce frost damage.

Gooseberries and currants tolerate a wide range of soils, but grow best on deep, well-drained loams with good moisture-holding capacity and large amounts of organic matter. While *Ribes* are less susceptible to root rot than raspberries and some other berry crops, they don't tolerate wet soils, and shouldn't be planted on poorly-drained sites.

For best fruit production, adjust soil pH levels to between 5.5 and 7.0. Have your soil tested by a reputable soil analytical laboratory in your area and follow their recommendations. When you have your soil test done, also have the lab test for calcium, total sodium, and total salt concentrations. *Ribes* should have at least 1000 parts per million (ppm) of calcium in the soil and can tolerate up to 15% calcium. Total salt and sodium concentrations shouldn't exceed 0.15% and 0.05%.

It's generally best to plant *Ribes* on a north-facing slope, if possible, to retard bloom and reduce heat stress. *Ribes* perform reasonably well in partial shade, although full sun is generally required for top production. In warmer climates, plant them in partial shade to reduce heat stress.

VARIETY SELECTION

Because of the wide variety of native and well-adapted introduced fruits, little currant or gooseberry breeding has been done in the United States since the early 1900s. Recently, interest in *Ribes* culture has increased in North America, and new varieties are being introduced from Europe and Canada.

Nomenclature in *Ribes* has been confused for centuries, and many varieties are known by several different names. While thousands of named varieties exist, relatively few of them are readily available in the United States, and not all of those are well-adapted to our growing conditions. The varieties listed in Tables 2 and 3 are

TABLE 2. Currant varieties for the home garden.

Name	Berry Size[1]	Berry Sweetness[2]	Bush Vigor[3]	Powdery Mildew[4]	Blister Rust[4]	Overall Rating[5]
Red Currants						
Cherry	S-M	V	V	R	R	G
Heros	M	M	V	M	R	G
Jonkheer van Tets	M	T	M	M	R	G
Minnesota 71	M-L	V	M	M	R	E
Perfection	L	V	V	M	R	E
Red Lake	S	M	M	M	R	G
Stevens No. 9	L	M	V	M	R	G
Wilder	S-M	M	M	R	R	G
White Currants						
Gloire des Sablons	L	M	V	M	R	F
Rosa Hollandische	M	M	V	R	R	F
White Imperial	S-M	V	V	R	R	E
White Currant 1301	M-L	M	M	R	R	G
Black Currants						
Ben Lomond	L	V	M	M	NR	G
Ben Sarek	L	V	M	R	NR	G
Black September	L-VL	M	V	R	NR	G
Boskoop Giant	M	M	V	M	NR	G
Consort	S-M	T	M	M	R	G
Crusader	M	M	M	M	R	G
Strata	VL	M	M	M	NR	G
Swedish Black	L	V	M	M	NR	G

[1] S = small, M = medium-sized, L = large, VL = very large

[2] T = tart, M = moderately sweet, V = very sweet

[3] M = moderately vigorous, V = very vigorous

[4] NR = not resistant, M = moderately resistant, R = resistant

[5] F = fair, G = good, E = excellent

TABLE 3. Gooseberry and jostaberry varieties for the home garden.

Name	Berry Size[1]	Berry Sweetness[2]	Bush Vigor[3]	Powdery Mildew[4]	Blister Rust[4]	Overall Rating[5]
Gooseberries						
Captivator	M-L	M	M	M	R	E
Chautauqua	M-L	M	L	M	R	F
Glendale	—[6]	–	V	M	R	F
Lepaa Red	S	V	L	R	R	G
Oregon (Champion)	S-M	M	M	M	R	G
Pixwell	S	M	M	R	R	E
Poorman	S	V	M	R	R	G
Speedwell	M	M	M	R	R	G
Welcome	–	T	M	–	R	G
Jostaberries						
Josta	L	M	V	M	R	G

[1] S = small, M = medium-sized, L = large, VL = very large

[2] T = tart, M = moderately sweet, V = very sweet

[3] L = low vigor, M = moderately vigorous, V = very vigorous

[4] NR = not resistant, M = moderately resistant, R = resistant

[5] F = fair, G = good, E = excellent

[6] – = reliable information not available

generally available from nurseries and widely adapted in North America. The performance of a particular variety, however, often varies from one growing region to another. The listed varieties are among the most reliable producers in North America. Many other varieties are available. If possible, make trial plantings of several varieties to determine which are best for your particular site.

Ordering

Before ordering gooseberries, currants, or jostaberries, make sure that you can legally import the plants and grow them on your site. A list of states that restrict *Ribes* importation or growing is given in

Table 1. Contact your state department of agriculture if you have any questions regarding these regulations.

Gooseberries, currants, and jostaberries are sold by nurseries as both bare root and container stock. Because of cost, most commercial growers purchase bare root plants. *Ribes* begin growing very early in the spring and are usually best dug and planted in the fall. For homeowners, containerized plants are more convenient and reliable than bare root stock. Whenever possible, order your plants in the spring for fall planting.

Bareroot and containerized gooseberry, currant, and jostaberry plants sold by nurseries are generally two years old, although you may occasionally get vigorous one-year old plants. Either is acceptable. The plants will have from one to several canes, which should be pencil thick and 12 to 18 inches (30 to 45 cm) tall. The root system should be dense and fibrous. Be especially careful to examine the roots of containerized plants. *Ribes* produce vigorous roots and can quickly become rootbound. Plants that have remained in containers too long will have many circling roots and practically no soil will be left in the containers. These plants dry out and wilt easily, and normally don't perform well when planted.

Plants should be healthy and free from pests and diseases. Reject those whose leaves or canes show dead, deformed areas or masses of white or brownish-gray fungus (powdery mildew). An important step in avoiding pest and disease problems is to start with clean, healthy stock in the first place.

POLLINATION

Currants, gooseberries, and jostaberries are pollinated by bees and other insects. Being among the earliest blooming fruits, they may suffer from poor pollination during cold, wet springs when bee activity is low. Red and white currants and gooseberries are generally self-fertile, which means that you only have to plant a single variety to get good fruit set. However, the degree of self-fertility varies according to the climate, and you may get better fruit set and size if you plant at least 2 varieties close together. Many black currant and jostaberry varieties are partly or entirely self-sterile, and you should normally plant two or more varieties close together. At the present time, the only jostaberry variety available in the United States is 'Josta.' The varieties 'Jostaki' and 'Jostine' should be available within the next several years.

CULTURE

Soil Preparation and Management

Be sure your site is well prepared before you plant. Correct soil nutrient imbalances, adjust soil pH, correct soil drainage problems, eliminate perennial weeds, and reduce pest and disease problems. *Ribes* normally don't grow well when planted onto a site immediately following heavy grass sod or well-established alfalfa. Buckwheat, barley, clover, vetch, oats, rape, beans, peas, and other garden vegetables are excellent rotation crops to precede gooseberries, currants, and jostaberries. You may also simply kill a turf, till up the area, and keep the soil bare for several months before planting *Ribes* to ensure good plant establishment and performance.

Planting

Because gooseberries, currants, and jostaberries begin growing very early in the spring, bare root stock is usually planted during mid- to late fall. Containerized stock can be planted any time from spring through fall, although the earlier you plant the better the bushes will establish before winter. Mulch summer and fall-planted bushes with straw, sawdust, or bark chips to reduce frost heaving.

Plant gooseberries 3 to 4 feet (90 to 120 cm) apart. Currants are more vigorous. Unless you're planning to train them to trellises or cordons, plant them 4 to 5 feet (120 to 150 cm) apart. Jostaberries are very vigorous, so plant them 6 to 8 feet (180 to 240 cm) apart.

If you're using bare root plants, carefully inspect the roots and use sharp pruners to remove any dead or diseased roots. Also prune off thick woody roots that are kinked, twisted around the crown, or point inward towards the crown. For container stock, cut through circling roots around the outside of the rootball using a sharp knife.

For bare root plants, dig a planting hole about 12 inches (30 cm) deep and 18 inches (46 cm) in diameter. Build a cone about 8 inches (20 cm) high in the bottom of the hole and spread the roots evenly over the cone, letting them drape toward the bottom of the hole. Set the bush somewhat higher than it grew in the nursery to allow for settling when you fill the hole. Prune any roots that won't fit into the hole. Fill the hole using only the soil that originally came out of it; don't add any soil amendments! Firm the soil over the roots using your hands, and settle it in place by soaking it several times with water. When you finish, the bush should be planted at the same depth as it grew in the nursery. Look for a change of bark color just above the root system. That shows the original planting depth. For containerized plants, dig a hole large enough to hold the root ball, and plant as described above. Be sure to water heavily immediately after planting.

Apply 4 to 6 inches (10 to 15 cm) of sawdust, bark chips, or other organic mulches around newly planted bushes to control weeds and keep the soil cool. You can use a liquid fertilizer solution to water in newly planted bushes (always follow label directions), or use dry fertilizer according to the rates given in Table 4.

Fertilization

Some people believe that currants and gooseberries don't need fertilizers, probably because neglected or abandoned bushes often

TABLE 4. Fertilization recommendations for *Ribes*.

Year	Amount of Fertilizer Per Bush Per Year						
	10-10-10	or	20-20-20	or	21-0-0	or	36-0-0
Planting	4 oz (115 g)		2 oz (55 g)		2 oz (55 g)		1 oz (40 g)
2	6 (170)		3 (85)		3 (85)		1.5 (55)
3	8 (230)		4 (115)		4 (115)		2 (75)
4*	10 (285)		5 (140)		5 (140)		3 (95)

*10-10-10 and 20-20-20 are common commercial formulations sold in garden centers. 21-0-0 is ammonium sulfate. 36-0-0 is ammonium nitrate.

survive for a long time. To ensure good vigor and production, however, fertilize your currants, gooseberries, and jostaberries every spring. Aside from nitrogen, potassium (potash) is the nutrient most often lacking for gooseberries and currants. Fertilizers that contain about 10% each of nitrogen (N), phosphorus (P), and potassium (K) (10-10-10) are commonly recommended, although a fertilizer ratio of 3:1:2 may provide a better nutrient balance. The amount of fertilizer to apply depends upon the age of the bushes. To prevent the fertilizer from leaching out of the root zone, it's generally best to put on half the fertilizer in the early spring just as the new growth is forming and half in the late spring when the fruit is beginning to develop. Fertilizer recommendations are given in Table 4. If you use uncomposted bark or sawdust as a mulch, apply an extra 0.25 pound (0.1 kg) of 10-10-10 fertilizer to each bush. If your soil and irrigation water are normally alkaline, use fertilizers that contain sulfate, such as ammonium sulfate, to acidify the soil over a period of time and to control soil pH.

Pruning and Training

Currants, gooseberries, and jostaberries are usually trained to individual, free-standing bushes or developed into hedgerows. Where space is limited, any of these crops can be trained on trellises or as cordons. Gooseberries may also be grafted to form "gooseberry trees," a complicated procedure. Trellis and cordon training are discussed in a later section of this chapter which also describes ways to save space in a garden or landscape.

Bush training requires the least time and skill. Because they have many canes, bushes also have the advantage of remaining healthy if one or a few canes are lost to pests or diseases. When the trunk of a cordon or tree-trained plant is killed, the entire plant dies. The negative feature of bushes is that canes may lie close to the ground where they interfere with cultivation and are exposed to high levels of moisture, pests, and disease organisms. Low-lying canes also make controlling weeds, insects, and diseases more difficult.

The goal of pruning *Ribes* is to keep canes young, ensure good air and light penetration, and remove diseased and damaged wood. Keeping the bushes open to light and air greatly reduces disease problems. Where growth is weak and few new canes develop, as is common in older plants, prune severely to stimulate new growth. Where growth is already vigorous, remove fewer canes. Use sharp tools to ensure clean cuts. High-quality bypass or hook-and-blade pruners generally give better results than anvil-type tools. You'll probably find long-handled loppers and leather gloves desirable when pruning gooseberries, due to the sharp spines on many varieties.

Except for cordon-trained plants, most *Ribes* are pruned while they are dormant during the late winter and early spring, but you can prune any time after the leaves have dropped in the fall. Fall pruning improves air circulation around dormant bushes during wet fall, winter, and spring months, and can decrease disease problems. In colder areas, you may want to wait until early spring to prune in order to identify and remove winter-damaged wood. Pruning currants, gooseberries, and jostaberries is similar to pruning blueberries (Figure 4). Remove crowded, injured, or drooping canes as close to the crown as possible. Canes normally aren't shortened or headed back unless they're damaged or diseased. If you must head back a cane, make the cut just above a side branch or strong bud. Be careful not to damage the spurs. Remove prunings from the garden and burn or otherwise dispose of them to reduce insect and disease problems.

With mature red and white currant, gooseberry, and jostaberry bushes, your goal should be to keep 3 strong, new canes per plant each year, and to remove an equal number of the oldest canes. In this system, mature plants have about 9 canes after pruning, 3 each of one-, two-, and three-year old wood. Wood that is 4 years old or older becomes unproductive and leggy. Black currants are more vigorous than other currants and gooseberries, and you normally leave about 10 to 12 vigorous canes per bush. If the bushes are very vigorous, leave a few more canes. About half of the canes left after pruning should be one-year old, with the remaining half being vigorous two-year old canes that have an abundance of one-year old shoots. Remove all canes that are more than 3 years old. The methods employed in pruning trellised bushes and cordons are shown in Figures 5 and 6.

FIGURE 4. Pruning currant, gooseberry, and jostaberry bushes. (A) At the time of planting, remove all two-year old and older wood. Leave 6 to 8 one-year old canes. (B) During the first dormant season, for all except black currants, remove all but 3 or 4 two-year old canes. Leave 4 or 5 vigorous one-year old canes. (C) During the second and subsequent dormant seasons, leave 3 or 4 each of one-, two-, and three-year old canes. Remove all canes 4 years old or older. For black currants, leave 4 to 6 canes each of one and two-year old wood and remove all wood 3 years old or older.

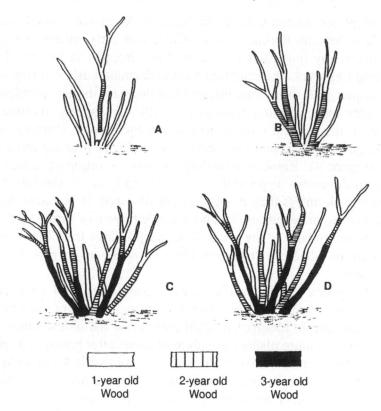

1-year old Wood   2-year old Wood   3-year old Wood

Watering

While gooseberries and currants favor cool, moist sites, they aren't particularly heavy water users. Approximately 1 inch of precipitation or irrigation water per week should normally suffice. On sandy or otherwise droughty soils, you may have to water more frequently. Be especially careful not to let containerized plants dry

FIGURE 5. Trellising for currants and gooseberries. (A) Plant currant and gooseberry bushes 3 to 4 feet (90 to 120 cm) apart and jostaberries 4 to 6 feet (120 to 180 cm) apart. Tie the canes to the bottom wire. (B-D) As side branches develop, tie them to the trellis wires. Prune using the methods given for free-standing bushes.

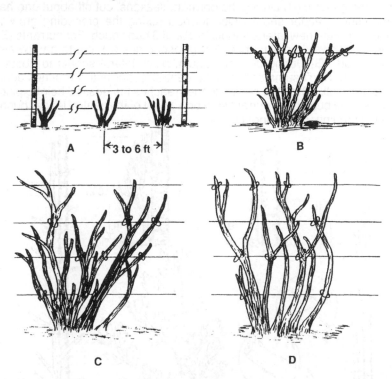

out and wilt. Any method of irrigation is suitable, but you can reduce disease problems by keeping the foliage dry. Rather than sprinkling the entire plant, apply the water directly to the soil around the base of each bush. If you use overhead sprinklers, apply the water early in the morning on warm, clear days to help the leaves and fruit dry quickly.

PROPAGATION

Currants, gooseberries, and jostaberries are relatively easy to propagate by layering or rooting cuttings. Except for breeding pur-

FIGURE 6. Training currants and gooseberries to a cordon. (A) Plant bushes on the south side of stakes. (B) Select a straight, upright, vigorous, one-year old cane. Cut off all other canes at the ground. Cut off about one-half of the selected cane, making the cut just above a strong bud. (C) When the central leader is 18 inches (45 cm) tall, remove all side branches within 6 inches (15 cm) of the ground. (D) During the dormant seasons, cut off about one-half of the central leader shoot which formed during the preceding growing season. Shorten new lateral shoots to about 3 buds each. For currants (E), cut to a downward-pointing bud. For gooseberries (F), cut to an upward-pointing bud. (G) During July or August, pinch lateral shoots to about 5 leaves each. Don't summer prune the central leader until it reaches about 6 feet (2 m) tall. When the central leader reaches the desired height, pinch off all but one bud on the current season's shoot of the central leader during summer pruning. (H) Mature cordon.

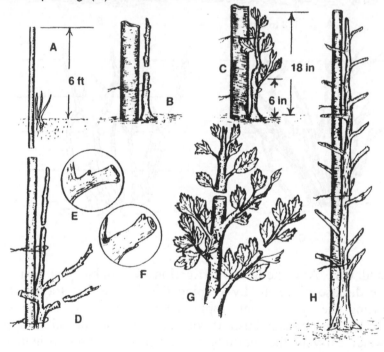

poses, the plants aren't grown from seed because each seedling has different characteristics. Although propagating gooseberries, currants, and jostaberries isn't difficult, it's easier and generally more satisfactory to purchase stock from commercial nurseries. Quality is a critical factor in propagating stock, and you must ensure that the plants are kept free from pests and diseases. Agricultural agencies inspect commercial nurseries and certify that they're taking steps to produce clean stock. Unless safeguards are taken, insects, mites, viruses, and diseases can be introduced into a garden or landscape on contaminated plants. Always propagate using material from vigorous, healthy, pest-free mother plants.

Currants, gooseberries, and jostaberries are typically propagated by layers or cuttings. It takes 1 to 2 seasons to grow a cutting or layer to planting size, depending upon the variety. Vigorous, easy-to-root varieties propagated in the spring may be ready to plant by the first fall. European gooseberries are often more difficult to propagate than other *Ribes* and may require 2 seasons to reach planting size.

Layering

Layering is usually done in the early spring. To produce a single rooted layer, bend a low-lying one-year old cane to the bottom of a shallow hole next to the mother plant and pin the cane into place with a U-shaped piece of wire, notched stick, or weight. Bend the cane sharply where it's pinned to the ground, and make a notch in or girdle the bark at that point to stimulate root formation. Fill the hole with garden soil and make a mound 4 to 6 inches (10 to 15 cm) high, leaving the rest of the cane exposed. After the plant becomes dormant in the fall, cut the rooted layer from the mother plant, taking care not to damage the new plant's roots. Keep as much soil around the roots as you can, and plant it into the garden as soon as possible.

A variation of simple layering, called stooling, can be used to produce larger amounts of stock. Start by cutting off all of the canes on an established bush 1 to 2 inches (2 to 5 cm) above the ground after the plants become dormant in the fall. In the following spring, after the new shoots have grown about 6 inches (15 cm) tall, mound sawdust over them, leaving several inches of the shoots exposed. Add more sawdust when the shoots are about 12 inches (30 cm) tall. Mounding with sawdust instead of soil allows you to more easily remove the rooted layers. When layering in a stool bed, you don't have to bend, pin, or wound the shoots before mounding. Roots form readily on the young shoots, and the layers can usually be cut from the stool bed that fall. New shoots will form the following season from the cane stubs and crowns in stool beds, which can remain productive for years. Simple layering and stooling are illustrated in Figure 7.

Cuttings

Both hardwood and softwood cuttings can provide good rooting success. Collect hardwood cuttings in late fall after leaf drop. Take 6 inch (15 cm) long cuttings from wood that grew during the past summer. Make a slanting cut on the base of the cuttings (closest to the mother plant) and a flat cut on the tip end to help identify the bases and tips of cuttings. Make the top cut about 0.25 inch (0.5 cm) above a bud. Bundle the cuttings together in groups of about 25 and store them buried upside down in the ground with at least 2 inches (5 cm) of soil or sand covering the butts of the cuttings until planting in the spring. Overwintering this way encourages callus to develop on the bottoms of the cuttings, which improves rooting. If you'd prefer, you can collect cuttings in mid- to late winter and store them in a refrigerator. Disinfect the cuttings by soaking them for ten minutes in a solution of 1 part household bleach in 9 parts of water. Then rinse thoroughly in running tap water for several minutes, wrap the cuttings in moist (not wet) paper towels, place them in a plastic bag, and store them in a refrigerator (not freezer). Disinfecting the cuttings helps prevent them from getting moldy during storage. Hardwood cuttings are normally stuck into pots for rooting in early spring, about the time of last frost.

Collect softwood or semi-hardwood cuttings from current season's growth during June or July. Young shoot tips tend to dry out rapidly and aren't as satisfactory as more mature tissues. Irrigate several days before collecting cuttings to ensure that the plants aren't drought-stressed, and collect the cuttings early in the morning. Make the cuttings 4 to 6 inches (10 to 15 cm) long, ensuring that each cutting has at least 4 buds. The cuttings should be about or slightly less than the thickness of a pencil. Leave 2 or 3 leaves at the tops of the cuttings and strip off all the other leaves. Cut off the outer one-half of the retained leaves to reduce moisture loss, and enclose the cuttings in plastic bags. It's best to begin rooting softwood cuttings immediately after you collect them, but you can store the cuttings for several days inside a plastic bag in a refrigerator. Don't allow softwood cuttings to freeze.

FIGURE 7. Propagating currants and gooseberries by layering. (A) Simple layering. While the plant is dormant in the spring, peg a single one-year old cane to the bottom of a 6 inch (15 cm) deep hole, tying the tip of the shoot to a stake above ground. Wound the cane by removing a strip of bark at the bend. Dig the layer in the fall or early winter after roots form. (B-E) Stool or mound layering. (B) Prune off all canes from dormant bushes, leaving 1 to 2 inch (2 to 4 cm) stubs above ground. (C) When new shoots are 6 inches (15 cm) tall, mound 3 inches (8 cm) of sawdust around the shoots. *Do not wound the shoots.* (D) when the shoots are 12 inches (30 cm) tall, add 3 to 5 inches (8 to 12 cm) more sawdust. (E) Dig the rooted layers as described above.

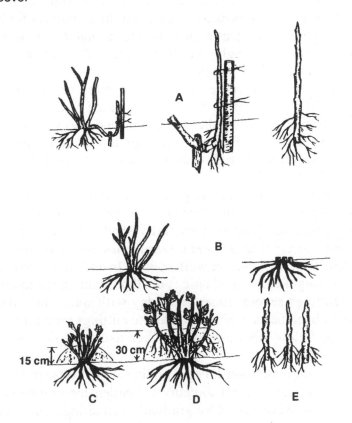

For best results, dip the bottoms of cuttings into rooting compound that contains from 1000 to 3000 parts ppm indole butyric acid (IBA). Hormones are especially useful for propagating hard-to-root European gooseberries. Commercial rooting hormone powders and mixes are generally available at your local garden store. Follow label directions.

Although any kind of flat or pot will work, sticking cuttings into individual #1 nursery pots (approximately 1 gallon) makes growing and transplanting easy. Wash used containers and sterilize them to kill disease-causing organisms by immersing the containers for 30 minutes in a solution of 1 part household bleach and 9 parts water. Rinse the containers thoroughly in clean, fresh water after sterilizing to remove traces of the bleach.

Fill the pots with a mixture of 50% peat moss or well-rotted compost and 50% perlite or sand. A 50:50 mixture of commercially-prepared potting soil and either perlite or sand also works well. Don't add any fertilizers. Plant the cuttings into the pots, making sure that the cuttings are right-side-up. Leave one or two buds exposed above the media.

Once the cuttings are stuck, irrigate them regularly and protect them from weeds, pests, and diseases. To prevent the cuttings from drying out, set them in a partially shaded location that's screened from direct sunlight and cover them with a tent of clear polyethylene film until they're well-rooted. In addition to irrigating and covering them with plastic, you'll probably have to mist softwood cuttings several times each day with water until they root. If mold becomes a problem, treat the cuttings with a fungicide. Always follow label directions when using fungicides and other pesticides.

Once the cuttings are rooted, begin adding slow-release or liquid fertilizers, according to label directions. Harden the cuttings off over about a two-week period by gradually increasing their exposure to sunlight before placing them in a full sun location. Gooseberries, currants, and jostaberries need to undergo a dormant period during the winter, so don't grow them indoors to extend their growing season.

PEST AND DISEASE CONTROL

Gooseberries and currants have the reputation of having few pests and diseases other than powdery mildew, blister rust, and currant or gooseberry fruitworm. This misconception comes mostly from the fact that these plants are usually grown in small, isolated plantings, which limits the spread of pests and diseases. If you have a lot of native or cultivated *Ribes* nearby, you'll probably have some pest and disease problems.

Few pesticides are registered in the United States to control pests and diseases on minor crops such as gooseberries and currants. Proper site and variety selection, keeping plantings clean, regular pruning, and other cultural practices are very important. If pests and diseases become serious problems, despite preventative cultural practices, consult with your local garden center or Cooperative Extension agricultural agent for help in selecting appropriate pesticides.

If they are registered for your area, dormant sprays of lime sulfur, Bordeaux mixes, and crop oil help prevent or reduce pest and disease problems. Apply a lime sulfur or Bordeaux spray in the fall or early winter after leaf drop to control fungal diseases. As the buds are just swelling in the spring, apply another sulfur spray to further reduce powdery mildew and other fungi. A dormant oil spray as the buds are swelling helps control mites, aphids, and other pests. Be careful when using Bordeaux mixes. These materials contain copper, which can build up to toxic levels in the soil if applied too frequently or in excessive amounts. Confine the spray to the canes only and only mix up as much spray material as you're going to use that day to avoid disposal problems.

Insect and Mite Pests

Cottony Maple Scale–Flattened, brown to yellowish-green larvae. Conspicuous cottony egg sacs of the mature form are two to three times longer than the scale body. Feed on foliage, reducing plant vigor and causing twig dieback. Apply pre-bloom dormant oil spray.

Currant Aphid–Found on red and white currants and gooseberries. The insect overwinters in the egg stage on the bark of new canes. The small, yellowish aphids begin to appear when leaf buds open in the spring. The aphids feed on the undersides of foliage, causing the leaves to redden and assume a distorted, cup shape. Honeydew on foliage and fruit is unsightly and makes fruit undesirable. Winged aphids develop in early summer and fly to other non-*Ribes* hosts. The aphids migrate back to the *Ribes* in the fall to mate and lay eggs. Control with insecticidal soaps or other pesticides. Protect populations of beneficial insects and mites by avoiding excessive or otherwise improper pesticide use.

Currant Borer—Adults are clear-winged, blue-black, wasp-like moths with yellow markings. Adults appear in late May or early June and deposit eggs on canes. Pale-yellow larvae tunnel in canes, weakening the plants and causing leaves to yellow and wilt during summer and autumn. Larvae overwinter in tunnels. Prune out and dispose of infested canes.

Currant Fruit Fly—One of the most serious pests of gooseberries and currants. Larvae bore out of infested berries, fall to the ground, and enter the soil in the summer. In the soil, they overwinter as pupae in brown cases about the size of wheat grains. Flies emerge from the soil in the spring and soon lay eggs in developing gooseberry or red, white, or black currant fruit. Adult flies frequently rest on fence rows, brush, and trees adjacent to currant and gooseberry plantings. Frequent, shallow cultivation under bushes helps expose and kill larvae. Some pesticides are registered for this pest.

Gooseberry Cambium Miner—Gray moths, 1/4 inch long, appear in June. Slender, semitransparent larvae mine up and down the cane cambium, stunting or killing the tops of new canes. Prune out and dispose of infested canes.

Gooseberry Fruitworm—A particularly serious pest of *Ribes*. Larvae bore into berries, and one larvae can destroy several berries. Larvae are green with a yellow tinge and darker stripes on their sides. Overwinter as a pupae under litter or underground, emerging as adult moths in the spring. Adults have ash-colored wings with dark markings, and a one-inch wingspread. No pesticides are registered. Generally controlled by currant fruit fly practices.

Imported Fruitworm—Full-grown larvae are about 1/2 inch long, greenish in color, and often have dark body spots, especially when partially grown. The larvae feed along leaf margins and can quickly defoliate plants when insects are numerous. Adults are black saw-flies with yellowish abdomens. Varieties vary widely in their susceptibility to this pest. Gooseberry-currant crosses seem to be especially susceptible. Pick off larvae by hand for small plantings. Some registered pesticides will control this pest.

San Jose Scale–Nearly circular and slightly convex scales, dark when small, but gray when fully developed. Attack leaves, flowers, shoots, and canes of all currants and gooseberries. Apply pre-bloom sprays of dormant oil.

Two Spotted Spider Mite–Overwinter as adults on weeds and debris at the base of bushes. They are about 1/50 inch long, have eight legs, and are light tan or greenish in color, with a dark spot on each side. Feeding reduces plant vigor and may cause leaves to turn brown and drop. Spider mites become especially troublesome when predator mites and insects are killed by improper pesticide use. Use pesticides only when absolutely necessary in order to protect predators. Reduce soil dust on and around bushes as much as possible.

Pests Other Than Insects and Mites

Deer–Potentially, the most serious non-insect pests are deer, which can severely damage young *Ribes* bushes by browsing on them. Even the sharp spines on gooseberries aren't a sure guard against hungry deer. In fact, a fence is the only sure defense against deer in your home garden. Deer quickly become used to and ignore scarecrows and other devices intended to frighten them away. Many commercial and home-made repellents are available, including bags of human hair, bars of soap, cat urine, blood meal, rotten eggs, and hot pepper sauce, but most remain effective for only a few days and few, if any, will repel deer that are very hungry. If you choose to use a repellent, make sure that it's legally registered for use on currants and gooseberries in your state, and always follow label directions. A chain link or wire mesh fence 4 feet (120 cm) high, with rows of wires or twine tied above the mesh to a height of 8 to 10 feet (2.5 to 3 m) normally gives adequate control. Use a mesh 6 inches (15 cm) or less square. Space the wires or twine about 10 inches (25 cm) apart.

Birds–Although currants and gooseberries are somewhat less attractive to birds than raspberries and blueberries because the fruit is more concealed by the foliage, birds can still damage a lot of fruit. For small gardens, covering the berry bushes with commercially available bird netting gives the easiest and most effective control. Be sure that there are no large holes in the netting and that it drapes all the way to the ground. Especially in windy areas, you might find it helpful to weigh the net down at the bottom or tie it around the bottoms of the bushes.

Nematodes–These microscopic worms can cause problems for gooseberries and currants. The most serious nematode for *Ribes* growers in the United States is the American nematode *Xiphinema americanum,* which appears to spread currant mosaic virus. Before planting, have your soil tested to identify nematode pests present and take steps to control them. Some pesticides, called nematicides, are available to control the pests. You may also obtain some control by solarizing the soil in the planting site. To do this, cover the moist soil with 2 layers of clear polyethylene film for the summer before planting. Solarization works best in warm climates that have frequent sunny days. Plant only clean, virus-free stock. Rogue out and dispose of infected plants.

Slugs–These pests attack the leaves and fruit of all *Ribes* and are especially active at night and during cool, wet weather. Overhead irrigation creates an ideal environment for slugs, which hide during the day in cracks in the soil and under debris. They damage plants by eating the leaves (usually between the veins) and fruit. Silvery-colored slime trails on damaged plant parts provide an easy diagnosis of slugs. Grass and weed borders, as well as debris within the field, all provide habitat for slugs. Good sanitation and weed control helps reduce slug populations. Prune bushes to open shapes to facilitate drying and air movement. Slugs typically infest perennial legumes. In areas where slugs are a problem, avoid using perennial clovers in garden cover crops. Pesticides are available which can effectively control slugs. Poison baits are the most popular control method and are available from most garden centers. Baits applied before the first fall rains will kill slugs before they lay eggs. An old but often effective home remedy for slugs is to make traps by placing shallow cups or pans filled with beer in a garden. The slugs are attracted to the smell of brewers yeast and drown after crawling into the container.

Diseases

Angular Leaf Spot—Infects the leaves of all currants and gooseberries, and can become serious. Large brown or gray spots with black specks in the centers develop on the leaves. Prune bushes to open shapes. Apply sulfur sprays.

Anthracnose Leaf Spot—Fungus overwinters on dead leaves and is especially serious during wet seasons. Causes very small leaf spots, resembling fly specks, and may result in yellowing and dropping of the leaves by midseason. It reduces the vigor, growth, and productivity of infected plants. Severely infected berries crack open and drop. Rake up and destroy leaves in fall. Apply sulfur sprays.

Armillaria Root Rot—Sometimes called shoestring fungus. Infects hundreds of plant species. Infected plants decline, produce no new growth, and gradually die. During the autumn, groups of honey-colored mushrooms or toadstools often appear at the base of infected plants. White felt-like mats or black shoestring-like strands of the fungus develop on roots. Dig up and dispose of infected plants. Don't replant in the same area for 2 or 3 years.

Blister Rust—Attacks both wild and cultivated gooseberries and currants. Black currants are most susceptible. Several black currant varieties are resistant to the disease. Five-needled pines are the alternate host for the fungus. Small, orange, cup-like spots develop on the undersides of infected currant or gooseberry leaves. In areas where five-needled pines grow, don't grow black currants or grow only rust-resistant varieties. Keep all gooseberry and currant plantings at least ½ mile (0.8 k) from the nearest susceptible pines.

Cane Blight or Wilt—Attacks the wood of currants, gooseberries, apples, roses, and other species. Red currants are the most susceptible *Ribes*. The disease can be very serious. Infected canes suddenly wilt and die during the summer, particularly as the fruit ripens. The wood and pith of diseased canes appear blackened. Later in the season, parallel rows of black, wart-like bodies appear on new and one-year-old canes. Prune out and destroy infected canes. Apply dormant sulfur or Bordeaux sprays.

Coral Spot or Dieback—Attacks the canes of all currants and gooseberries. Old and neglected bushes are very susceptible. Occasional branches may show wilting soon after leaves develop, or symptoms may not appear until the fruit begins to ripen. The bark of infected canes may be covered with pink pustules. Infection occurs through dead snags, branches, or pruning stubs. Prune out and destroy infected canes. Apply dormant sulfur or Bordeaux sprays.

Currant Mosaic Virus–Can weaken and kill infected red and white currants. The primary symptoms are mottling of the leaves. The American nematode has been identified in spreading the disease. Don't plant where nematode exists. Use healthy stock. Dig up and destroy infected bushes.

Dieback and Fruit Rot–The botrytis fungus attacks the canes, leaves, flowers, and fruits on currants and gooseberries, especially during rainy weather and wet conditions. The fungus lives on dead branches, fruit, and foliage, moving to healthy tissues in splashing rain and irrigation water. Infected plants are weakened and can be killed. Stems can be girdled. Leaves become yellowish-gray at the margins and may fall prematurely. Infected berries turn brown and rot. Infected tissues can be covered with a soft gray felt-like mass containing black spots. Prune bushes to open shapes. Irrigate early in the morning. Apply sulfur sprays just before and after bloom. Plant in sunny locations.

Gooseberry Rust–Also called cluster cup rust, this fungus attacks the leaves, stems, and fruit of gooseberries. Sedges are the alternate host for this rust organism. Seldom a problem, but can become serious during periods of wet weather. Outbreaks usually occur on wild gooseberries or in abandoned gardens. Infected leaves are thickened in areas where the cups will later appear. The spots become red and later the cups break through the undersides of the leaves. Similar symptoms occur on the stems, petioles, and fruit. Eliminate sedges in and around the garden. Destroy or discard prunings.

Powdery Mildew–Both European and American forms of this disease affect *Ribes*. The fungi overwinter on gooseberry and currant twigs, attacking shoots, leaves, and fruit. Although European powdery mildew is not a particularly serious disease, American gooseberry powdery mildew is extremely serious on susceptible varieties. The fungi appear as white powdery growths on the surfaces of leaves, green shoots, and fruits. Infected plants are often stunted and severely affected plants can be killed. As the fruit matures, the mildew changes to a dark brown, felt-like coating that makes the berries inedible. Affected leaves develop dead, brown, dry areas, become deformed, and dry out. Plant only resistant varieties. Prune bushes to open shapes. Plant in sunny locations. Apply sulfur sprays just before and after bloom, and during the summer, as needed.

Septoria Leaf Spot–Infects the leaves of currants and gooseberries. Most destructive in the Mississippi Valley, but reported throughout continental United States, Alaska, and Canada. Seldom causes severe crop loss. Defoliation reduces the vitality of bushes and exposes fruit to sunscald. Small, brown spots resembling anthracnose appear about June. The spots enlarge and the central area becomes light-colored with brown borders. Small, black specks are scattered over the spots. The diseased leaves, especially on currants, turn yellow and drop. Prune bushes to open shapes. Rake up and discard leaves in fall. Usually controlled by powdery mildew control programs.

Weeds

Currants, gooseberries, and jostaberries don't compete well with other plants, and getting weeds under control is critical for healthy bushes and good yields. If you're planting into an area covered with grass sod, kill the grass early the year before planting and either keep the ground bare or plant to buckwheat, barley, clover, vetch, oats, rape, beans, peas, or other garden vegetables.

Mulches control annual weeds well but aren't very effective against perennials, such as Canada thistle and quackgrass. If you plan to use a mulch, eliminate perennial weeds before you plant the bushes. A contact herbicide, such as glyphosphate, gives the easiest and most thorough control. Organic growers will have to depend upon rotation crops, solarization, and repeated weeding to prevent the perennial weeds from getting more than an inch or so tall.

To control weeds after planting, mulch under the dripline of the bushes. Sawdust, bark chips, and clean, weed-free straw (not hay!) are all commonly-used. Most compost provides excellent conditions for weed seed germination and growth, and may not give adequate weed control when used as a mulch. Straw and similar mulches can also increase mouse girdling damage, especially during the winter.

Landscape weed control fabrics can be used around gooseberries and currants that are grown as individual bushes, and are generally effective, especially against annual weeds. Many fabrics won't control perennial weeds such as yellow nutsedge and quackgrass. If you want to apply decorative mulch over a weed control fabric, use coarse bark chips or gravel, rather than a fine-grained mulch, to reduce germination of weed seeds on top of the fabric. Some people like to use black polyethylene plastic film as a weed barrier. Because the film is impermeable, it creates problems with watering and fertilization, and I don't recommend using it for weed control in *Ribes*.

After perennial weeds and grasses are eliminated, hand cultivation provides good weed control, but keep it shallow to prevent damaging the roots. Cultivating under the bushes also helps control currant fruit fly and gooseberry fruitworm by exposing and killing the pupae. Be cautious when using a rototiller between the rows for weed control. Some people recommend deep cultivation around the bushes in hot, dry areas to develop deep root systems. Cultivation more than about 1 inch (2.5 cm) deep, however, damages the shallow roots. Deep cultivation also brings new weed seeds to the surface where they can germinate, and excessive cultivation compacts and damages the structure of soil, decreases soil organic matter, and increases soil erosion. The goal in weeding is to cut the weeds off just at the soil surface.

HARVESTING

Currants ripen over about a two-week period, depending on the variety. You may want to harvest in two pickings to get the berries at their peak, although berries will remain on a bush without falling or becoming over-ripe for a week or more after ripening. Berries normally begin ripening in July, depending upon your location, but cold weather during spring and early summer can delay ripening. Red and white currants are glossy and attractive, while black currants, gooseberries, and jostaberries appear dull because of chlorophyll in their skin.

Red and white currants have delicate skins and are harvested by pinching off entire clusters or strigs at their bases. The berries may be stripped from the stems just before processing. Because black currants are firmer than red or white currants, they can be picked individually or stripped from clusters right in the garden. Since fruit for juices and jellies is strained before processing, you don't need to remove stems before crushing berries for these uses. For fresh use, pick into 1/2 pint (250 ml) containers and leave the clusters intact. Baskets and flats designed for raspberries work well for fresh currants. If you're going to process the fruit, pick it into 1, 2, or 4 quart (1, 2, or 4 liter) containers.

Currants intended for jellies and other preserves are often picked slightly underripe because the fruit pectin content is highest at this time. In the United States, however, the tendency has been to pick both currant and gooseberry fruits far too early because of the misconception that ripe fruit won't jell properly during cooking. Unripe currants and gooseberries are very unpalatable. Ripe berries taste much better than green ones and can be used effectively for processing. For fresh use, allow the berries to fully ripen before picking. When ripe, currants will be soft and flavorful, and will be fully colored with no trace of green on the stem end. Overripe fruit shrivels and becomes mushy.

Gooseberries begin ripening at about the same time or slightly later than currants, and are harvested over a four to six week period. For best flavor, gooseberries should be ripe or just slightly underripe when picked for processing. They are generally picked about 3 times each season. During the first picking, harvest about 1/3 of the fruit, leaving the remainder evenly distributed on all the canes. Thinning the fruit increases its size by increasing food reserves available to the remaining berries. During the second picking, strip berries from low-lying canes and those canes in the center of the bush, leaving the best quality fruit to ripen on well-exposed outside branches. Berries from the first two pickings are normally slightly underripe and are used for processing. Those from the third picking are fully ripe and excellent for fresh use.

Gooseberries are very firm and resist bruising when green, but soften as they ripen. Be more careful when picking ripe than underripe fruit. Because of their size, firmness, and the presence of thorns on many varieties, gooseberry harvesting tends to be rather "rough," resulting in a lot of leaves and other debris being collected into the picking containers. Gooseberries for processing are harvested by stripping them from the bushes with heavy gloves or cranberry scoops. Berries for fresh or dessert use are picked individually by hand into pint or quart (500 ml or 1 liter) containers. Processing berries can be picked into 1, 2, or 4 quart (1, 2, or 4 liter) baskets or flats. To remove debris, pour the fruit gently from one container to another in front of a fan. Jostaberries are harvested like gooseberries.

Be very careful not to crush the berries when picking, and don't collect damaged berries, even for processing. The juice from crushed berries increases the development of fruit rot, and a few crushed and rotting berries can ruin all of the fruit in a container. Water on harvested berries also increases fruit rot problems. Pick the fruit only when it's dry, and don't wash the berries until just before you use them.

USE IN THE LANDSCAPE

Currants are gaining in popularity as easily-managed landscape shrubs. Their early growth and bloom, small size, and attractive, edible fruit make them prime candidates for edible landscapes. Their thorns and small size make gooseberries effective as low barrier and border plants along property boundaries, but watch out for the thorns if you or your neighbors have small children. For landscape use, select the most disease-resistant varieties available. Be careful not to use varieties susceptible to blister rust if there are any five-needled pines within about 1/2 mile (0.8 k) of your planting. Powdery mildew is the most serious problem in landscape plantings. Again, disease resistance is critical.

Both currants and gooseberries lend themselves especially well to raised beds mulched with bark or other organic materials. You can plant them either in full sun or partial shade locations. Because they're very vigorous and tend to develop rank growth, jostaberries are less adaptable to landscape use than currants or gooseberries, but might be useful along borders or to screen fences and outbuildings.

In addition to edible currants, several ornamental currants are used in landscapes. *Ribes odoratum,* also known as clove currant, can be useful in providing a border, particularly if you want a fragrant garden in the early spring. Clove currants are susceptible to blister rust. 'Crandall' is probably the best known variety of this species. Clove currants are generally not grown for fruit, but when fully ripe the large, black berries of 'Crandall' have a mild, pleasant flavor.

Alpine currant, *R. alpinum,* is a European native often used as a hedge plant, particularly in semi-shade locations. Recommended for hardiness zones 2 to 7, alpine currant normally grows 3 to 6 feet (90 to 200 cm) tall, but can reach 10 feet (3 m) if left unpruned. It has deep, bright green leaves and is among the first shrubs to leaf out in the spring. The flowers are inconspicuous, and the fruit isn't edible. 'Aureum' and 'Green Mound' are common varieties.

Winter currant, *R. sanguineum,* is adapted to cool, moist coastal areas of the Pacific Northwest in hardiness zones 5 to 7. The 6 to 10 foot (2 to 3 m) tall shrub produces long racemes of white, pink, or red flowers in early spring. Plant this species in partial shade to reduce leaf fading and scorching. 'King Edward VII,' a 5 to 6 foot (1.5 to 2 m) tall bush with red flowers, and 'Brocklebankii' (pink flowers and yellow leaves) are probably the most common varieties.

SPACE SAVING IDEAS

Trellises

Currant and gooseberry bushes can easily be trained to fan shapes that are supported by trellis wires. Their vigorous, rank growth can make trellising jostaberries somewhat more difficult. Trellising conserves space, simplifies weed, pest, and disease control, and creates an attractive wall. Start a trellis system by setting individual currant and gooseberry plants every 3 to 4 feet (90 to 120 cm) next to a trellis which has 3 to 5 horizontal wires spaced about 12 inches (30 cm) apart. Jostaberries should be spaced 4 to 6 feet (120 to 200 cm) apart. The height of the trellis depends on the vigor of the bush. Gooseberries can be trained on smaller trellises than currants or jostaberries. As side branches develop on the canes, tie them to the wires. Keep the ties loose and check them regularly to ensure that they don't cut into or girdle the branches. When pruning, follow the guidelines given for bushes (Figure 4). A typical trellised bush is illustrated in Figure 6.

Cordons

Cordon-trained plants have single, straight trunks about 6 feet (2 m) tall that are supported on stakes. Any currant, gooseberry, or jostaberry can be trained to a cordon. As with trellises, cordon-trained plants take up less space than bushes, are attractive, and simplify picking. Cordon training reduces fruit yields, however, and can increase sunscald and berry cracking on sunny sites. Another problem with cordons is that damage to the trunk can kill an entire plant.

To establish a cordon system, set out individual plants in rows running north and south. The rows should be at least 6 feet (2 m) apart. When the plants to be cordoned are especially vigorous (black currants and jostaberries), space the rows farther apart. Within the rows, set gooseberries, red currants, and white currants 2 to 3 feet (60 to 90 cm) apart. Space black currants 3 to 4 feet (90 to 120 cm) and jostaberries 4 to 6 feet (120 to 200 cm) apart. The steps in developing a cordon are shown in Figure 7. By following these steps, you should be able to keep a healthy cordon about 6 feet (2 m) tall for many years.

Growing Grapes in the Home Garden

M. Ahmed Ahmedullah

INTRODUCTION

Grapes are easy to grow in the home garden. They are well adapted to different soils and climates, and can be trained to a wall, fence or trellis, thus requiring much less space than fruit trees. Their cultural requirements are minimal. They are a favorite small fruit, relished for fresh fruit, or juice and popular for jam and jelly, for drying as raisins and for making home wines. The grape berries may be red, blue, white, greenish-yellow, bronze, purple or black. Each possesses distinctive aroma, flavor and quality.

CLASSIFICATION AND ORIGIN

The hundreds of varieties of grapes grown in North America can be grouped into three distinct types.

1. The European grape, also called the Old World Grape (*Vitis vinifera* L) is the basis of the very large California wine, table grape, and raisin industry. It is also grown in Europe and much of the rest of the world. Many hybrids of *vinifera* have been developed for

M. Ahmed Ahmedullah, Department of Horticulture and Landscape Architecture, Washington State University, Pullman, WA 99164-6414.

[Haworth co-indexing entry note]: "Growing Grapes in the Home Garden." Ahmedullah, M. Ahmed. Co-published simultaneously in *Journal of Small Fruit & Viticulture* (Food Products Press, an imprint of The Haworth Press, Inc.) Vol. 4, No. 3/4, 1996, pp. 143-188; and: *Small Fruits in the Home Garden* (ed: Robert E. Gough, and E. Barclay Poling) Food Products Press, an imprint of The Haworth Press, Inc., 1996, pp. 143-188. Single or multiple copies of this article are available from The Haworth Document Delivery Service [1-800-342-9678, 9:00 a.m. - 5:00 p.m. (EST). E-mail address: getinfo@haworth.com].

© 1996 by The Haworth Press, Inc. All rights reserved.

143

earliness and disease/pest resistance. The *vinifera* varieties include table grapes for eating out of hand, wine grapes, and raisin grapes which are dried into raisins. Most *vinifera* varieties are not winter hardy. They can be grown in states with long, warm-to-hot dry summers and cool winters. Because of their extreme susceptibility to fungus diseases and insect pests, *viniferas* are not recommended in regions with humid summers. They are grown in California, Arizona, New Mexico and in the best locations in Oregon and Washington states. Hardy varieties can be grown in certain mid-western and eastern states if special precautions are taken to protect them from winter cold injury.

2. American grapes, also called Fox grapes (*Vitis labrusca* L), are the principal "New World" species. They are comparatively more winter hardy and have a typical "foxy" flavor. There are hundreds of varieties in this group. Hybrids have been developed that combine the delicate flavor and quality of *vinifera* with the hardiness of *labrusca*. This group can be further divided into those varieties used for wine making, those used for fresh fruit and those used for processing into juice, jams and jellies. An example of a blue grape in this group is 'Concord' which is processed into juice, jams and jellies.

French Hybrids

French hybrid grapes are varieties introduced by French grape breeders who developed them by crossing the European *vinifera* varieties with native American wild species grown in North America. These carry the name of breeders and/or originators and in most cases the seedling number. Examples include 'Seibel' 10868, 'Seyve-Villard' 5-276, 'Baco' 1, 'Ravat' and 'Kuhlmann.' Grapes in this group are primarily used for wine making.

3. The Muscadine grapes, *Vitis rotundifolia* Michaux were the first American species to be cultivated. Muscadines grow wild and are also cultivated in the southern United States. They are not cold hardy, but have resistance to many diseases. The oldest and best known muscadine variety is 'Scuppernong.' The term "Scuppernong" is sometimes also used to refer to all bronze colored muscadine varieties. Muscadines can also be divided into wine types, table types and those grown for juice, jams and jellies. The better

varieties in this group were developed in breeding programs in the southern United States.

Structure of Grapevine

The grapevine consists of the above ground parts, the trunk, the arms, the canes (mature shoots) and the leaves; and the below-ground parts–the root system (Figure 1). Familiarity with the specialized terminology used in grape production will help you understand the culture a little better. An extensive glossary is included at the end of this chapter. The grapevine needs some support from either a fence, wall or trellis to grow vertically. Familiarity with some terms given later in these pages will help one to understand the discussion on training and pruning.

Nutritive Value of Grapes

Grape berries contain 70 to 80% water and 15 to 25% sugars which consist of glucose and fructose. The berries also contain vitamin A, B-complex, and small quantities of vitamin C. The vitamin content of *labrusca* grapes is not as high as *vinifera* grapes.

Adaptation

Decide which type of grape grows best in your area. European or *Vinifera* grapes require milder climates such as those found in California, Arizona, New Mexico and parts of Oregon and Washington states. The hardy *labrusca* types can be grown in most states. Remember, muscadines can only be grown in the south and southeast.

Within each type, you should also decide if the grapes will be utilized for eating fresh, wine making, juice, jam, jelly, or for drying into raisins.

VARIETY SELECTION

Most people like to grow grapes in home gardens for eating fresh out-of-hand, though some might use them for wine and preserves.

FIGURE 1. Diagrammatic illustration showing important structures of a grapevine.

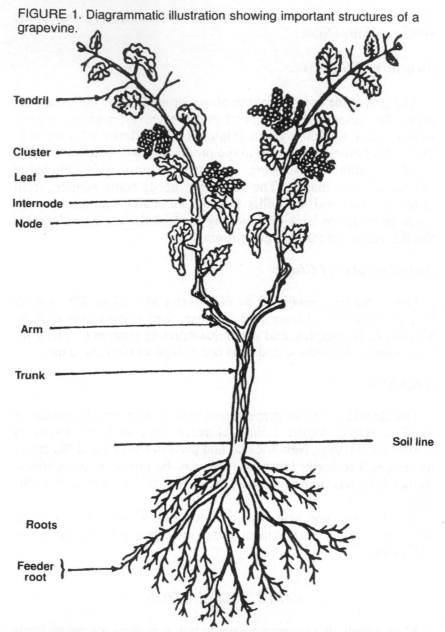

Tendril

Cluster

Leaf

Internode

Node

Arm

Trunk

Soil line

Roots

Feeder
root

Source: Grape Growing by R. J. Weaver. 1976. Reprinted by permission of John Wiley & Sons, Inc., New York.

When you have decided how you will use your grapes, select the appropriate varieties. Varieties suitable for growing in different regions in the United States (Figure 2) are suggested in the following. The maturity dates given may be somewhat different from the actual maturity date in your home garden because of difference in climate. You should also consult your cooperative extension agent or request literature from the state extension service in your area. Select varieties which are early, midseason, and late-ripening. This will ensure a continuous supply of grapes with diversity of flavor and a continuous supply throughout the growing season. All varieties suggested are self fruitful, meaning that only a single variety

FIGURE 2. Grape growing regions in United States.

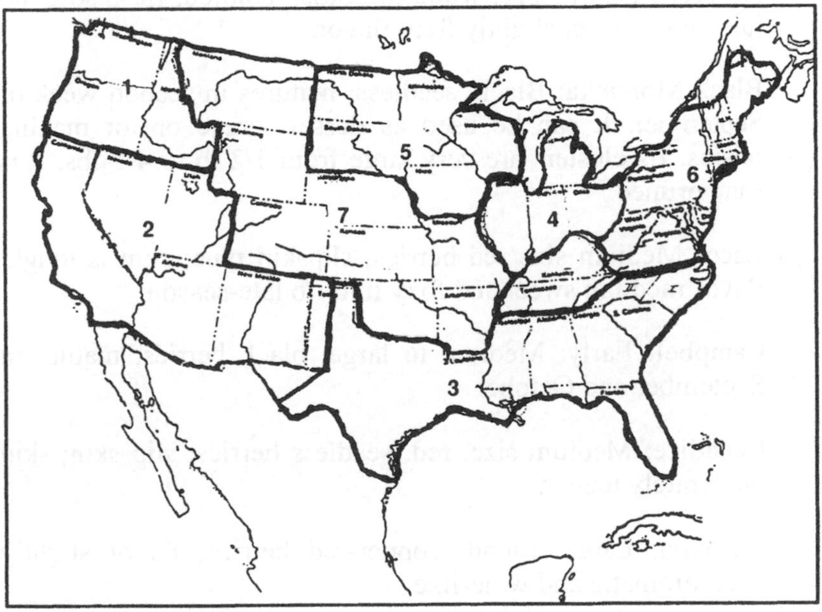

1. Northeast
2. Southwest
3. Southeast
4. North Central
5. Central and Northeastern
6. Northeastern
7. Central

needs to be planted. Of course you can plant two to four vines of several varieties or if several rows are planted, devote one-half row to each variety. A single mature vine will produce 20-40 lbs of fruit each year. Two to four vines will give 40 to 160 lbs of fruit or three to six gallons of juice.

Region 1. Northwest Region
(Oregon, Washington and Idaho, British Columbia, Canada)

Only winter hardy and early ripening varieties can be grown here because of the short growing season and low winter temperatures. Here are some varieties to try. Most varieties are spur-pruned except where indicated.

> Aurore: Medium size, white-to-golden colored; flesh soft and juicy with sweet slightly foxy flavor.

> Black Monukka: Black, seedless, matures in second week of September. It can be used as dessert grape or for making raisins. Its clusters are very large from 1/2 lb to 4-5 lbs. It is cane pruned.

> Caco: Medium size red berries, slipskin type; skin is tough; flavor medium sweet and foxy mid- to late-season.

> Campbell Early: Medium to large, black berries, mature in September and October.

> Canadice: Medium size, red, seedless berries; slip skin; skin moderately tender.

> Catawba: Large, round, copper-red berries; flavor slightly foxy, aromatic and wine-like.

> Delaware: Small, round, red berries; flavor is good.

> Einset Seedless: Medium size, oval-shaped, red seedless; not a slipskin type; flavor fruity, mild *labrusca*-type, with strawberry character.

Fiesta: Small to medium, white seedless berries; flesh firm; skin is tender and flavor is good.

Flame Seedless: Small to medium size, round, red seedless berries mature in late July to mid-August. Flesh is firm and crisp. Has excellent flavor.

Golden Muscat: Medium to large, yellow-green berries; flesh juicy, soft, sweet; has muscat flavor; skin tough.

Himrod: Medium size, white seedless berries, mature early in September. The clusters are large, loose and irregular. It is excellent as dessert grape.

Interlaken Seedless: Small, white seedless berries, mature early, a month earlier than Concord. Clusters are of medium size and tight.

Lakemont: Fruit small to medium, seedless, yellow-green berries; non slipskin; fruit juicy, sweet.

Moored: Medium, red berries, early; flesh has foxy flavor.

Niagara: Large, greenish-white; flavor is sweet with foxy characteristics.

Suffolk Red: Medium size, red, seedless berries; early to mid-season, ripens a few weeks before Concord.

Utah Giant: Large, round, blotchy red fruit; flesh is firm and crisp; good as dessert grape.

Vannesa Seedless: Medium, oval, red, seedless; flesh is crisp with good flavor; early to midseason; matures 10-14 days before Concord.

Region 2 Southwest Region
(California, Arizona, New Mexico, Western Texas and Utah)

Because of the long growing season and ideal climate, you can grow many *vinifera,* and *labrusca* varieties in California's home gardens. Some early varieties can also be grown in home gardens in the best areas of Arizona, New Mexico and Utah. Try these varieties.

Beauty Seedless: Medium size, black, seedless berries mature in mid-July.

Black Monukka: Medium to large, black, seedless berries can be used both as dessert grape and for making raisins. Its clusters are very large from 1/2 lb to 4-5 lbs. It is cane pruned. The berries ripen before Thompson Seedless.

Black Rose: Seeded, black berries mature in mid- to late-season.

Blush Seedless: Red seedless berries mature in late August to early September.

Cardinal: Large, red, seeded berries mature early in September.

Centennial Seedless: Medium-large, seedless, white berries ripen in late July.

Christmas Rose: Tear-drop shaped, red, seeded berries mature in mid-September.

Dawn Seedless: Medium to large, white, seedless berries, are tough skinned. They mature in late July.

Emperor: Medium to large, seeded, red to purple berries mature in late September to early October.

Exotic: Large, black and seeded berries mature in early August.

Flame Seedless: Medium size, red, seedless berries mature in late July.

Muscat of Alexandria: Medium size, seeded, white berries are muscat flavored. They ripen in mid-September.

Perlette: Small to medium, white, seedless berries mature in mid-July.

Queen: Large, red, seeded berries mature in September.

Red Globe: Very large, bright pinkish-red, seeded berries mature in mid-September.

Ribier: Large, black, seeded berries mature in late August.

Ruby Seedless: Medium to large, elongated, seedless berries mature in mid- to late August.

Thompson Seedless: This most popular table grape variety produces medium, round, white seedless berries which mature in mid-August to September. The vines should be head trained and cane pruned leaving 10-14 buds on each cane. With each cane retain a two-bud renewal spur for getting a cane for next year.

Region 3 Southeast Region
(Alabama, Georgia, Louisiana, Florida, North Carolina, South Carolina, Mississippi and Eastern Texas)

North Carolina is divided in two subregions. In subregion I, varieties Carlos and Scuppernong can be grown. In subregion II, American, French hybrids, some hardy *viniferas* can be grown.

The following grape varieties are recommended for the homeowner in this region.

Albemarle: Muscadine grape, medium size, blue-black berries mature in third week of September. It has good flavor; recommended for planting in northern and southern part of Louisiana. (Not recommended for south Florida.)

Black Fry: Muscadine grape, large, grayed purple, thick skinned berries, mature in the third week of September.

Blue Lake: Muscadine grape, small to medium, blue berries mature in June and July. The fruit has an aromatic and spicy flavor.

Cascade: French-American hybrid, small to medium, round, black berries, mature early, in the first to second week of October.

Chancellor: French-American hybrid, medium size, blue berries with thin skin, mature midseason to late in the first to second week of October.

Chelois: French-American hybrid, small to medium, round, reddish-blue berries mature early midseason, shortly before Concord, third week of September to first week of October.

Cowart: Muscadine grape, large, bluish-black berries, mature in mid-September. It is very productive.

Daytona: Muscadine grape, light green to pink berries, mature in August.

Dixie: Muscadine grape, medium size, whitish bronze berries mature in September.

Dixieland: Muscadine grape, medium size, bronze to light-red berries mature in August.

Fry: Muscadine grape, large to very large berries, skin thick; matures third week of September to first week of October. Not suitable for subregion I in North Carolina.

Golden Isles: Muscadine grape, medium size, bronze colored berries; flavor very good; mature in midseason; ripening not uniform; more suitable for wine.

Granny Val: Muscadine grape, medium to large, bronze berries, quality good; mature in midseason.

Higgins: Muscadine grape, large to very large bronze berries with thick skin; mature in midseason to late.

Ison: Muscadine grape, large, round, bronze, thick skinned berries mature in the third week of September.

Janet: Muscadine grape, large, bronze berries; flavor excellent; mature midseason.

Jumbo: Muscadine grape, large, black, round berries with thick skin, mature in first to second week of September.

Pam: Muscadine grape, large, bronze berries, mature in mid-season.

Nesbitt: Muscadine grape, large, round, black berries mature in the first week of September to first week of October; has a very extended maturity period.

Noble: Muscadine grape, medium size, black, round berries, mature early-midseason; good for wine.

Seyval: French-American hybrid, small to medium, round to oval, white to yellowish-white berries; mature early midseason, two weeks before Concord.

Southland: Muscadine grape, medium to large, purplish-black berries, mature in second week of September; very productive, producing 40-50 lbs of fruit per vine.

Stover: Muscadine grape, small, light green berries mature in July.

Summit: Muscadine grape, medium size, bronze berries; very sweet; skin thick; mature in midseason.

Supreme: Muscadine grape, medium size, colored berries; mature in midseason.

Tara: Muscadine grape, large bronze berries; mature in midseason.

Triumph: Muscadine grape, large, bronze berries mature in July and August.

Vidal: French-American hybrid, small greenish-white berries, mature in the third week of September to first week of October.

Villard blanc: French-American hybrid, large, oval berries, white to golden-white berries mature late midseason to late.

Region 4 North Central Region
(Ohio, Indiana, Michigan, Illinois)

Because of the cold winters in this region, it is difficult to grow the *vinifera* grapes. If you want to try *viniferas* for making wine, grow 'Chardonnay,' 'White Riesling,' 'Cabernet franc' or 'Cabernet Sauvignon' grafted onto phylloxera resistant rootstock. The following *labrusca* and hybrid varieties are recommended without reservation.

Alden: Reddish-black, large fruited berries mature in late September to early October.

Aurore: Small to medium, white to golden-white berries mature very early in the first week of August.

Baco noir: Small, blue-black berries, mature in the second week of September, two weeks before Concord.

Bath: Medium size, blue-black berries, mature in the fourth week of September to first week of October.

Buffalo: Medium size, reddish-black berries, mature in the first to second week of October.

Caco: Large, red berries, mature in late midseason.

Canadice: Medium size, red, seedless berries mature in the second week of October.

Catawba: Medium size, purplish-red berries, mature late midseason in the fourth week of October.

Cayuga White: Medium to large, white berries mature in the third week of September at the same time as Concord.

Chambourcin: Medium size, blue berries mature in the first to third week of October.

Chancellor: Small, blue berries mature in October.

Chardonnel: Medium size, amber berries mature in the second week of October.

Chelois: Medium size, blue berries mature in the third week of September to second week of October.

Concord: Blue-black, medium size berries, mature midseason in October. It can be used for juice, jams, jellies and also as a fresh fruit. Fruit has typical 'foxy' flavor.

Concord Seedless: Small to medium, blue-black berries, mature in early midseason, in September to early October; one week earlier than Concord. Mature unevenly which will be an advantage for the homeowner to get a continuous crop.

De Chaunac: Small to medium, blue berries, mature in the third week of September to second week of October.

Delaware: Small, pink to light red berries, mature midseason in the first to second week of October. Very sweet.

Einset Seedless: Medium, bright red, seedless berries, mature in first to third week of September.

Fredonia: Bluish-black, large berries, mature early in September. Has flavor like that of Concord.

Foch: Small, black berries mature early in August to September.

Golden Muscat: Medium size, white berries, mature in late midseason. Has *vinifera*-like fruit characteristics.

Himrod: Small to medium, white to greenish-yellow, seedless berries, mature early to early-midseason in August. Quality very good.

Horizon: Medium size, seedless, white berries, mature in the third to fourth week of September.

Interlaken Seedless: Medium, white, seedless berries, mature in early September.

Lakemont: Medium size, yellowish-green, seedless berries, mature in September.

Mars: Medium size, blue, seedless berries, mature in the second week of September.

Reliance: Medium to large, grayish red, seedless berries, mature in the first to second week of September.

Romulus: Small to medium, white to yellowish-green, seedless berries, mature early, in the last week of August to second week of September.

Melody: Medium size, light yellow-green berries mature in the fourth week of September to first week of October.

Monticello: Medium size, blue-black berries, mature early. Quality very good.

Moore's Early: Medium size, blue-black berries mature in September.

New York Muscat: Small to medium, reddish-black berries, mature in the third week of September to second week of October.

Niagara: Medium to large, whitish-green berries, mature in September and October. Quality is good.

Ontario: Medium to large, white to golden-yellow berries, mature in the second to third week of August to first week of September.

Price: Medium size, red berries mature in the first week of September.

Seneca: Medium to large, yellowish-white berries, mature in the second week of September.

Seyval: Small to medium, yellowish-white berries, mature in the third week of August to second week of September; two weeks before Concord.

Schuyler: Small to medium, blue-black berries mature in the second to fourth week of September.

Steuben: Medium to large, blue berries mature in the first to third week of October.

Urbana: Medium size, thick skinned, red berries, mature late in October.

Van Buren: Medium size, bluish-black berries mature in the last week of August to second week of September.

Vannesa: Medium size, bright red, seedless berries, mature in the second to fourth week of September.

Vidal blanc: Small, greenish-white berries mature in the third week of September to second week of October.

Vignoles: Small to medium, white berries mature in the third week of September.

Villard noir: Medium size, blue-black berries mature in the first to second week of October.

Many of the grape varieties listed above can only be grown in the most favorable climatic zones in all the north central states.

Region 5 Central and Northeastern Region
(Minnesota, Wisconsin, Iowa, North and South Dakota)

Here are few varieties to try in this cold area.

Agawam: Medium-large, purplish-red berries, mature late, in the second week of October.

Beta: Small to medium size, black berries mature in the second to third week of September; very hardy.

Clinton: Small, black berries, mature in the second week of October; hardy.

Edelwise: Medium size, green berries, mature in the second week of September; hardy.

Kay Gray: Medium size, yellow-green berries, mature in second to third week of August in Wisconsin; hardy.

Swenson's Red: Medium-large, red berries, mature in the second to fourth week of September; hardy.

Worden: Large, black berries, mature in the second to fourth week of September; hardy.

Region 6 Northeastern Region
(New York, Pennsylvania, Maryland, Delaware, New Jersey, Connecticut, Rhode Island, Massachusetts, New Hampshire, Vermont, Maine, Virginia, West Virginia and Ontario, Canada)

The *vinifera* varieties are not winter hardy here. Therefore they are not recommended to be planted. If you want to try them, however, plant them only in a very protected location. Two varieties which might make it are 'White Riesling' and 'Chardonnay.'

The following *labrusca* and French-American hybrids can be grown.

Athens: Medium to large, reddish-black berries, mature early, in the first to second week of September; three weeks before Concord.

Aurore: Small to medium, white to golden-white, berries mature very early in the first week of August.

Baco noir: Small, blue-black berries, mature in midseason, in the second week of September, two weeks before Concord.

Bath: Medium size, blue-black berries, mature in midseason, in the fourth week of September to first week of October.

Brighton: Medium to large, light-red berries, mature in midseason, in the second to fourth week of September.

Buffalo: Medium size, reddish-black berries, mature midseason, in the first to second week of October.

Caco: Large, light-red berries, mature in late midseason.

Cascade: Small to medium, black berries; mature in the first to second week of September; early.

Catawba: Medium size, purplish-red berries, mature late midseason in the third week of October.

Cayuga White: Medium to large, white berries, mature in midseason, in the third week of September, at the same time as Concord.

Chancellor: Small, blue berries mature late, in the second week of October.

Chelois: Medium size, blue berries, mature midseason, in the third week of September to first week of October.

Concord: Medium size, blue-black berries, mature midseason, in the second week of October. It can be used for juice, jams, jellies and also as a fresh fruit. Fruit has typical 'foxy' flavor.

Concord seedless: Small, blue-black berries; seedless, mature earlier than Concord; excellent for preserves; ripening uneven which is an advantage for the homeowner; has good flavor.

De Chaunac: Small to medium, blue berries mature midseason, in the third week of September to second week of October.

Delaware: Small, pink to light-red berries, mature midseason, in the first to second week of October. Berries are very sweet.

Dutchess: Medium size, pale yellowish-green berries, mature medium-late, in the first to third week of October.

Fredonia: Large, bluish-black berries, mature early midseason, in September.

Foch: Small, black berries, mature early, in August to September.

Golden Muscat: Medium size, white berries, mature late mid-season. Berries have *vinifera*-like characteristics.

Himrod: Medium size, white seedless berries, mature early, in the first week of September. The clusters are large, loose and irregular. It is excellent for dessert.

Interlaken Seedless: Small, white, seedless berries, mature a month earlier than Concord.

Isabella: Medium-large, blue berries, mature medium-late, in the third week of September to first week of October.

Lakemont: Medium size, white, seedless berries, medium early, mature one to two weeks before Concord.

Leon Millot: Small, red berries, mature early, in the first to third week of September.

Niagara: Medium to large, whitish-green berries, mature mid-season, in September and October. Quality is good.

Portland: Medium to large, whitish green, berries mature medium early, in the second week of September.

Rosette: Small to medium, blue-black berries, mature medium late, in the fourth week of September to first week of October.

Rougeon: Small to medium, red berries, mature late midseason, in the first to second week of October.

Rumulus: Small to medium, seedless, white to yellowish-green berries, mature early, in the last week of August to second week of September.

Seyval: Small to medium, yellowish-white berries, mature midseason, in the third week of August to second week of September; two weeks before Concord.

Sheridon: Large, black berries, mature medium late, in the second to third week of October.

Steuben: Medium to large, blue berries, mature in the first to second week of October.

Suffolk Red: Large, red, seedless berries, mature medium early.

Urbana: Medium, thick skinned, red berries, mature late, in the second week of October.

Vignoles: Small to medium, white berries, mature midseason, in the third week of September.

Villard blanc: Small, greenish-white berries, mature late, in the third week of September to second week of October.

Region 7 Central Region
(Arkansas, Colorado, Missouri, Nebraska, Oklahoma, Kansas, Southern Illinois and Tennessee)

This is a large region covering different types of climates. *Labrusca* and French American hybrids are grown in this region. Muscadine grapes, may not be sufficiently winter hardy in some areas except the southern portion of this region. The following varieties may be tried.

Alden: Large, reddish-black berries, mature in late September to early October.

Alwood: Medium, blue berries, slipskin type, moderately foxy; mature in the second week of September.

Aurore: Small to medium, white to golden-white berries, mature very early, in the second week of August to second week of September.

Buffalo: Medium size, reddish-black berries, mature midseason, in the first to second week of October.

Canadice: Medium size, pink, seedless berries, mature in early midseason.

Carlos: Muscadine grape, medium to large, bronze berries, mature in the fourth week of August to second week of September. Suitable primarily for southern portion of central region.

Catawba: Medium size, red, oval to spherical berries, mature late, in the fourth week of October, 2 weeks later than Concord.

Chancellor: Medium size, round, blue-black berries, mature in late-midseason, in the second week of October.

Chelois: Small to medium, round, blue berries, mature midseason to late; in the first to third week of October.

Concord: Medium to large, globular, blue-black berries, mature in late midseason.

Cowart: Muscadine grape, large, blue-black berries, mature in mid-September. It is very productive. Suitable primarily for the southern portion of central region.

Cynthiana: Small to medium, black berries, spherical, late.

De Chaunac: Small to medium, round, blue berries, mature early-midseason.

Delaware: Small, short ovate to round, pink to light red berries, mature late midseason.

Foch: Small, black berries, mature early, in August to September. Flavor like that of Concord.

Fredonia: Large, bluish-black berries, lack the typical Concord flavor, mature early-midseason.

Golden Muscat: Medium to large, yellow-green berries; flesh juicy, soft, sweet; has muscat flavor; skin tough; mature early midseason.

Himrod: Medium size, white seedless berries, mature early, in the first week of September. The clusters are large, loose and irregular. It is excellent for dessert.

Lakemont: Medium size, white, seedless berries, mature early in mid-August.

Magnolia: Muscadine grape, medium to large bronze berries, mature in the second to fourth week of September. Suitable primarily for the southern portion of central region.

McCampbell: Medium, blue-black berries, mature early in the first to second week of September.

Mars: Medium size, seedless, blue berries, mature in early to mid-August.

Moored: Medium size, round, red berries, mature 3 weeks before Concord.

Niagara: Medium to large, thin-skinned, white berries, mature in midseason.

Reliance: Medium size, pink, seedless berries, mature in early midseason.

Saturn: Very large, red, seedless berries, mature in the second week of September.

Seyval Blanc: Small to medium, round to oval, white berries, mature early-midseason.

Steuben: Medium to large, spherical, blue berries, mature late-midseason.

Venus: Large, seedless, blue-black berries, mature early in mid-August.

Verdelet: Small to medium, pale-white to amber-white berries, mature in early midseason, in the third week of August to second week of September.

Villard Blanc: Small, greenish-white berries, mature late mid-season, in the third week of September to second week of October.

Villard noir: Medium size, blue-black berries, mature late, in the first to second week of October.

FACTORS TO CONSIDER IN PLANTING
THE HOME VINEYARD

Site

Choose the sunniest spot in your garden with good air and water drainage. Good air drainage is usually important for grapes because fungal diseases are quite prevalent. A southern exposure usually gets better sunlight, so the grapes will ripen earlier there. A northern or eastern exposure usually makes the crop ripen later because of low sunlight. Avoid areas which get direct cold wind.

Length of the Growing Season

For growing grapes, the length of the growing season (the period between the last spring frost to first fall frost) should be at least 155 to 165 days, with longer periods being more desirable. Choose a variety that will mature in your area.

OBTAINING THE PLANTS

Grapevines are easy to propagate by hardwood cuttings, but you will be better off purchasing your rooted plants from a reliable nursery which sells certified virus-free vines. A vineyard is a long term investment even for a small homeowner, and it will be false economy to buy cheaper plants from an unreliable nursery or to grow your own. In addition purchasing the one-year-old rooted hardwood cuttings has the added advantage of saving the year's time that is required to obtain a rooted plant from a cutting. Two-year-old vines are preferred to one-year-old vines. You should make sure that the two-year-old vines have a strong stem and root system. Softwood propagated plants are also available from some nurseries. These are shallow-rooted and are not recommended in regions subject to areas experiencing winter temperatures of $-5°F$ or lower. Grapevines propagated from seeds will not be true to type.

PREPARATION OF PLANTS
FOR PLANTING AND PLANTING THE VINES

The site selected for planting the vines should be free of weeds and other vegetation. If the area is small, it can be hand cultivated to remove weeds, especially perennial weeds. If the area is large, it should be ploughed to improve the physical condition of the soil.

The rooted vine bought from a reputable nursery will be dormant in late winter or early spring. In many cases they will have many roots 9 to 18 inches long and several buds on the top. Trim the roots to 6 to 9 inches keeping only the strongest and best roots. Never bend the roots to make them fit into a hole that is too small. Such roots will not grow well and are likely to make the plant shallow rooted. Such vines are often damaged during the cold winters.

Planting

Do not allow the plants to dry out when they are received. If you are planting several vines, mark their location with a chain or string and wooden pegs. This will assure the rows will be straight. Dig holes with a shovel, 2 to 3 feet deep and 1 1/2 to 2 feet wide. Fill the hole halfway with the loose soil and plant the vine. Do not allow the plants to dry out when they are received. If the conditions are not favorable for immediate planting, store the vines in a cool place such as the basement and loosely cover them with plastic to minimize the loss of moisture. The roots can be soaked in water overnight if they appear dry. Try to plant the vines as soon as holes are dug. In any event do not store for more than a week.

Place some top soil or compost at the bottom of the hole, filling it about half way. Place the vine in the hole, spreading the roots well and fill the hole with the soil up to the point where roots and stem meet. Then tamp the soil with your feet as the hole is filled so that no air pockets remain in the vicinity of roots. The vine should be thoroughly watered immediately after the hole is filled.

Distance Between the Vines

Plant most vines 6 feet apart. More vigorous varieties should be planted 8 feet apart. For Muscadines 10 feet is best in the row. If several rows are planted, the distance between the rows could be 5 to 7 feet. There should be enough distance between the plants in the row so that the adjoining vines will not intermingle and the roots of vines will not overcrowd each other.

BEST TIME TO PLANT THE VINES

In most areas of the United States spring planting is preferable to fall planting. Spring planting allows the vine to become well established before the onset of winter months. In areas which do not experience severe winters, like Texas, Louisiana and Florida, the vines can also be planted in the fall.

CARE OF THE YOUNG VINES

The grapevine will not grow a straight trunk without support. Stake the vine or tie it to a trellis to keep the shoots growing straight (Figure 3a). By late spring the young vine should be growing vigorously, producing many shoots. Retain these as their leaves will produce food to help develop a strong root and shoot system. Select the strongest shoot in midspring to become the trunk (Figure 3b). Tie this shoot to the stake or trellis loosely with a string to encourage straight growth. Select a 4 to 6 feet long stake and plant it close to the vine. Several tyings will be required as the shoot grows in length. Control weeds growing near the trunk either by hoeing or with herbicides registered for use around the home. Many of the commercial herbicides cannot be used legally in home vineyards.

TRAINING OF YOUNG VINES

Continue to train and prune your vines each year. The exact method, however, will depend upon the type of vine support system you have selected.

FIGURE 3a. Stake the vine and tie it to the support.

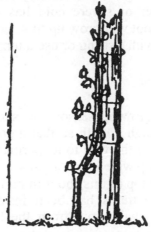

Source: Training and Trellising Grapes in Washington. 1990. M. Ahmedullah, Extension Bulletin 637, Cooperative Extension, Washington State University, Pullman, WA 99164.

FIGURE 3b. Select the strongest shoot which becomes the trunk.

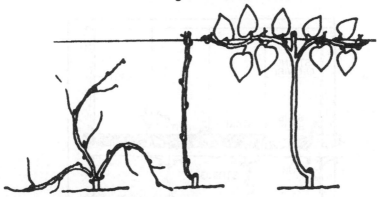

Source: Training and Trellising Grapes in Washington. 1990. M. Ahmedullah, Extension Bulletin 637, Cooperative Extension, Washington State University, Pullman, WA 99164.

First Growing Season

At the end of the first growing season, when the vine is dormant, prune to one to two buds on the strongest canes. Remove all other shoots by pruning as close to the trunk as possible. In the northern states, because of the danger of cold, delay pruning until the time when you think danger of severe cold has passed. During the spring, the strongest shoot will grow up to 4 feet or more. It should be tied to the top wire with a string or use a temporary stake.

Second Growing Season

During the second growing season the shoot selected for the trunk should make enough growth so that a good beginning could be made towards training the vine to its permanent form. The cane that was tied up to the top wire in the last season should have grown to 4 to 5 feet. Prune the top of this shoot in midspring, a few inches below the wire (if the trellis has been installed) to encourage branching. New shoots will grow from the buds below the cut (Figure 4).

FIGURE 4. Training the vine in the first and second year after planting.

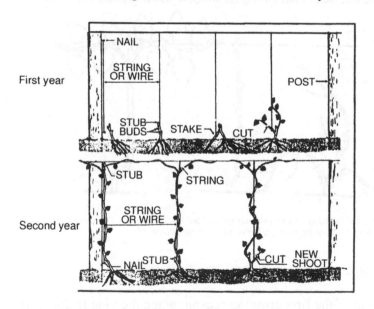

Source: Training and Trellising Grapes in Washington. 1990. M. Ahmedullah, Extension Bulletin 637, Cooperative Extension, Washington State University, Pullman, WA 99164.

Training the Vine to a Bilateral Training System

The following discussion assumes that you are training your vine to a bilateral training system. This system is easy to maintain once established. For other systems, discussed under trellising systems, the details would differ.

Third Growing Season

Select two shoots, each growing in opposite directions from the trunk for a single wire bilateral training system. Tie these to the wire with one shoot extending in each direction from the main trunk (Figure 5) and allow them to grow during the season. Since the vine is still too young to grow a full crop, cut back these shoots so that only 5 to 7 buds remain on the canes on either side of the trunk. Tie the cane to the wire loosely using baler twine or other soft material.

FIGURE 5. Growth of vine in the third growing season.

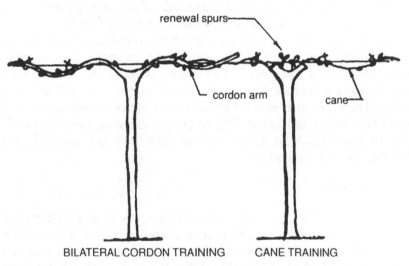

BILATERAL CORDON TRAINING   CANE TRAINING

Third Year

Source: Trellising and Training Grapes in Washington 1990. M. Ahmedullah, Extension Bulletin 637, Cooperative Extension, Washington State University, Pullman, WA 99164.

Give the canes one and a half twist around the wire to prevent it from slipping. If a wooden support is provided for the trunk, tie the trunk loosely to the support. The trunk can also be tied to the wire.

Several shoots will arise from the 5 to 7 buds left on the canes. Each shoot can give 1 to 3 clusters of grapes depending upon the variety. The vine will bear a small crop in the third year. Try not to take a big crop in the first season the vine bears fruit, because this could weaken the vine in the long run. Remove extra flowers from the vine to reduce the crop. Twenty clusters is considered a good crop for the first season.

As winter approaches and the temperature drops, the vine will gradually lose its leaves. The shoots by this time will become hardened canes. This will take place during November, December or January, depending upon the season and the region.

VINE SUPPORT SYSTEMS–TRELLIS CONSTRUCTION

You must decide what type of vine support system to use. There are several types of trellises available. Most vertical trellises for vineyards in the eastern United States are of the same general type: two to three wires, one above the other, stretched tightly on firmly set posts. Two wires are adequate for most training systems. If you are growing only a few vines you might construct an overhead arbor, also called "pergola." The other common choices for vine support system include a one wire trellis and a two or three wire trellis. There are also several other newer types of trellises which could be easily constructed. The overhead arbor and one and two-wire trellises which are more common are described and only the sketches of others are given.

Overhead Arbor

For an arbor, four pressure treated posts, 10-14 feet long with a top diameter of 3-4 inches, should be buried at least 2-3 feet in the ground. The posts should be 10-12 feet from each other in a square or rectangular fashion. Securely anchor the posts with concrete and rock. Using nails or metal brackets, connect the top of the four posts with strong wooden poles. Install additional poles every 3 feet to form a lattice on which the shoots will grow.

Plant the vine either in the center or on one side of the arbor and allow a single trunk to grow straight from the ground alongside a permanent support post. Pinch all laterals on the lower part of the trunk, to favor the strong growth. When the vine has reached the top of the arbor, it is pinched-in, or cut-back so as to make it grow shoots and spread out, from the head of the vine. When the shoots grow horizontally, leave spurs at intervals of 24 inches, from which to renew the wood from year to year. Leave one cane, 3 or 4 feet long on each spur at pruning time. Shoots springing from these will cover intermediate spaces soon after growth starts. After 3 growing seasons, the arbor will be complete (Figure 6). The clusters hang down from a canopy of leaves over the top.

Training Systems for American and French Hybrid Grapes

If you are planning to grow American or French hybrid type varieties, use the systems discussed below.

FIGURE 6. Vine trained to an overhead arbor.

Four-Arm Kniffen

This system is characterized by the four short arms from which the fruiting shoots arise. The arms, two on each side of the trunk, are developed from the trunk extensions and renewal spurs (Figure 7). These provide fruiting wood for the following year. Canes are tied horizontally along the wires. Cane length may vary from 8 to 12 buds depending upon the variety and cane vigor. Prune away all surplus buds each year. Select and train the four canes arising from the top of the vine on the top wire as they will be more productive than those arising from the lower portion on the trunk.

FIGURE 7. Four-arm kniffen system.

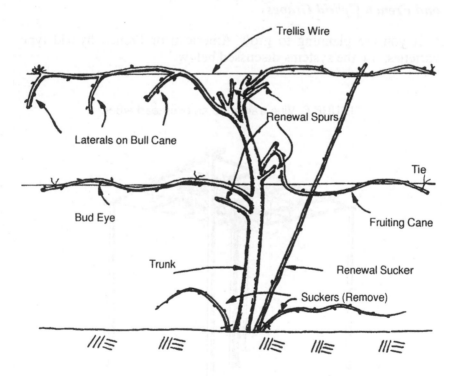

Source: Grape Training and Pruning by C.C. Schaller published by Cooperative Extension Service, University of Arkansas.

Other Training Systems

Other training systems used for grapes are: six-arm kniffen, the Umbrella kniffen, Hudson River Umbrella and the modified Keuka High Renewal. Four-arm Kniffen (Figure 7) is easy and most practical for the homeowner and is described. Other systems are not discussed here but their sketches are given (Figure 8).

Trellising Systems

The following trellising systems can be used for vigorous growing *vinifera, labrusca* or French hybrid varieties.

Single-Wire Vertical Trellis

Place seven-foot wooden posts with a minimum top diameter of 3 to 4 inches, vertically, 18 to 21 feet apart (every 3 vines), and bury at least 2 feet in the ground for adequate support. Use poured concrete or rocks for firm support. Place a single # 9 wire across the top of the post, 5 feet above the ground (Figure 9) and secure it with several U-shaped nails. Cordon-train the grapes along the wire in both directions. This system is least expensive, suitable for 'Concord' and other *labrusca*-type grapes and for *vinifera* varieties. You can also construct this trellis close to a wall. You should select a wall or site which gets the maximum sunlight.

Two-Wire Vertical Trellis

Place seven-foot long wooden posts with a minimum top diameter of 3 to 4 inches vertically, 18 to 21 feet apart (every 3 vines), bury at least 2 feet in the ground. Use poured concrete or rocks for strong support. Place the top wire across the top of the post with a second wire placed 16 to 30 inches lower (Figure 10). Secure the wires firmly in place using several U-shaped nails.

Train the vines on a bilateral cordon fashion on the top wire. Use the bottom wire for training two canes on either side of the trunk. Use the No. 11 wire on the top of the post, 5 feet above the ground level. This is sometimes called foliage wire. Use a second wire, No.

FIGURE 8. Other training systems for grapes.

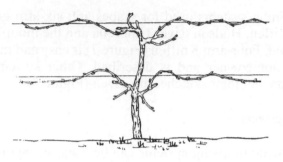

Four-Arm Kniffin[a]

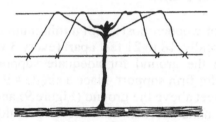

Umbrella Kniffin[b]

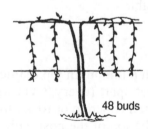

48 buds

Hudson River Umbrella[c]

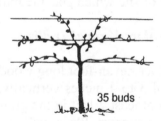

35 buds

Modified Keuka High Renewal[c]

Source: [a]Growing Grapes in West Virginia, Circular 117 (Revised), March 1981, N. Carl Hardin, Tara L. Auxt, and Steven H. Blizzard, West Virginia University Cooperative Extension Service Bulletin. [b]Growing Grapes in Arkansas, Cooperative Extension Service, Bulletin 563, University of Arkansas. [c]Cultural Practices for Commercial Vineyards, Bulletin 111, T. D. Jordan, R. M. Pool, T. J. Zabadal, and J. P. Tomkins, Cooperative Extension, New York State College of Agriculture and Life Sciences, at Cornell University.

FIGURE 9. Single-wire vertical trellis.

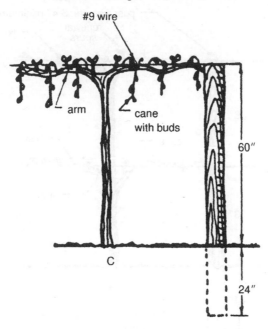

Source: Training and Trellising Grapes in Washington. 1990. M. Ahmedullah, Extension Bulletin 637, Cooperative Extension, Washington State University, Pullman, WA 99164.

9 wire, 16 inch below the top wire and 44 inches above the ground level for cordon support. Secure both wires firmly in place with U-shaped nails.

Use this system for *vinifera, labrusca* or French hybrid varieties.

Three-Wire "T" Trellis

The top wires of "T" trellis are parallel. They are separated and supported by a cross arm attached to the post. This provides for better utilization of sunlight. Select 8-foot long posts with 2 feet buried in the soil; 6 feet above the soil level. In this 3-wire system, locate the center wire 6 feet above the ground (Figure 11). Across the top of the post (6 feet above ground) attach a 3 to 4-foot wooden or metal cross arm. Attach the 3 parallel wires to each end of the cross arm to support the fruiting canes. Use a #12 wire for the center and two #9 wires at each end of the cross arm. The main advantage

FIGURE 10. Two-wire trellis.

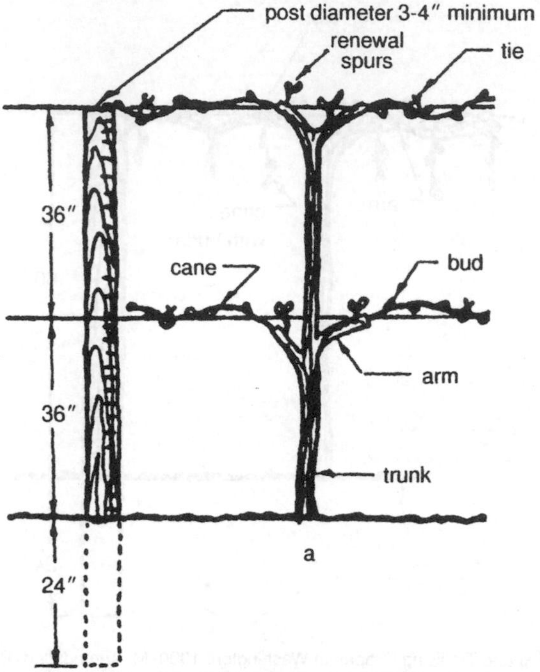

Source: Training and Trellising Grapes in Washington. 1990. M. Ahmedullah, Extension Bulletin 637, Cooperative Extension, Washington State University, Pullman, WA 99164.

of the "T" trellis is greater exposure of leaves to sunlight. The cost of installation is higher than the single wire or 2-wire vertical systems. For *vinifera* grapes, the lower or main wire is placed 4 1/2 feet (54 inches) above the ground to carry the main arms of the vine.

The new trellising systems, namely (1) Gable trellis, (2) Sloping Arm trellis, and (3) Wye "Y" trellis (Figure 12) are less common. Once properly installed they look very nice. Their sketches are given here (Figure 12) without description so that the homeowner has a choice of growing his vines on one of the new systems which are aesthetically more appealing than the conventional systems discussed above.

The advantage of these systems over the bilateral cordon system is that more leaves are exposed to the sun, thus increasing the food manufacturing capacity of the vine and improving the fruit quality.

FIGURE 11. "T" trellis.

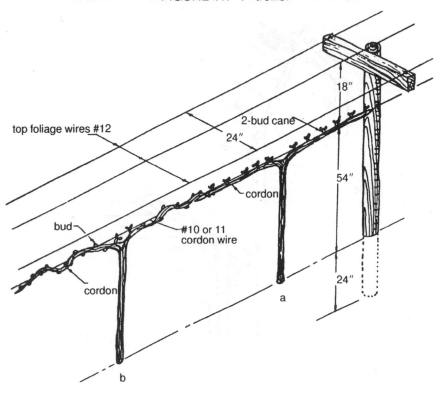

Training Systems for the Vinifera Varieties

Head-Trained Spur Pruned

Training the vine for the first two years for this system is the same as discussed earlier for other systems. Head-trained vines require short-term support. When they have established a stout, strong trunk, 3 to 4 inches in diameter they stand on their own without support (Figure 13).

To form the head of a head-trained vine, you should have a single straight cane tied vertically to the stake, with several laterals growing on it. Cut the cane at the node, above the level at which the head is desired and tie the cane to the wire or stake. This will induce branching. Allow the vine to grow uninterrupted. Next season, the

FIGURE 12. Gable, sloping arm and wye "Y" trellises.

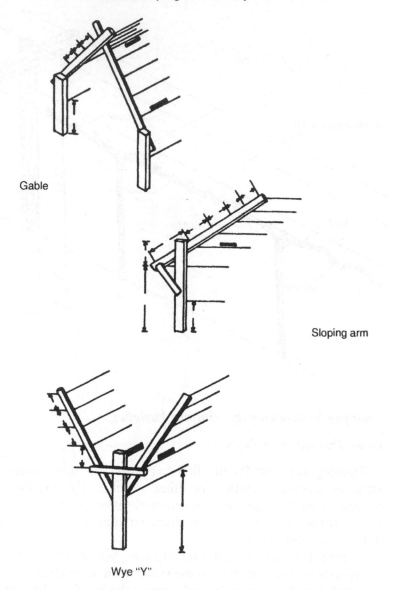

Gable

Sloping arm

Wye "Y"

Source: M. Ahmedullah. 1983. Less Common Trellising and Training Systems. Proceedings of Pacific Northwest Grape Shortcourse. Eds. Watson and Ahmedullah, Cooperative Extension, Washington State University, Pullman, WA 99164.

FIGURE 13

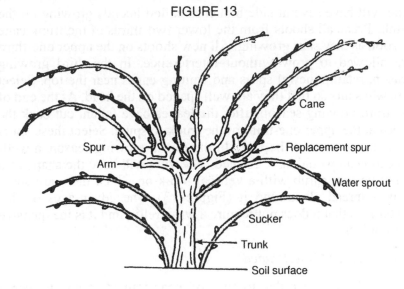

Cane

Spur →
Arm →

Replacement spur

Water sprout

Sucker

Trunk

Soil surface

Head—training
spur—pruning

vine will have several side branches called laterals growing on the trunk. Prune all shoots from the lower two-thirds of the trunk cane as soon as they start growing. All new shoots on the upper one-third are allowed to grow without interference. In the third growing season, retain renewal spurs and fruiting canes near the top. Select renewal spurs so that they are well spaced on the head. At the end of the third growing season, after the leaves have fallen, cut back the canes at the upper one-third of the trunk to spurs. Select these spurs close to the head of the vine. In the fourth growing season, a well-spaced head would have formed. The vine will have the shape of a small upright shrub with a vertical trunk on which there are spurs (arms) spread all around it (Figure 13). The advantages of this system are that it does not require a wire trellis, and it is inexpensive to establish.

Head-Trained Cane Pruned

The shape is similar to that of head-trained vines discussed above, except that the head may be fan-shaped in the plane of the trellis (Figure 14). A wire trellis is required for this system. Retain only two or three arms on each side of the head. At winter pruning, retain fruiting canes with 8 to 15 buds depending upon the variety for fruit production. For every cane retained, also retain a two-bud spur. This spur will give you the canes for next year. The advantage of cane pruning is that a full crop is obtained with varieties which have few or no fruitful buds at the base of the cane as in Thompson Seedless.

Cordon-Trained Spur Pruned

The vines of the horizontal cordon system have no definite head. A wire trellis is required. On the cordon, arms are distributed over greater parts of the trunk at intervals of 8 to 12 inches. At pruning time in winter, retain 2 to 4 bud spurs on each arm depending upon the variety (Figure 15).

PRUNING

Pruning refers to removing the unwanted plant parts. The grapevine produces many more buds on its canes than are required for a

FIGURE 14

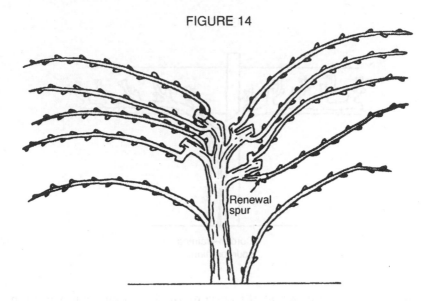

Renewal
spur

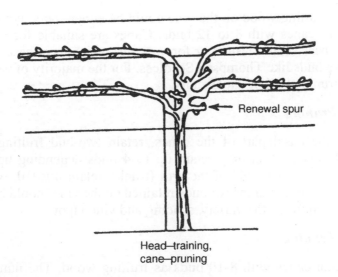

→ Renewal spur

Head—training,
cane—pruning

FIGURE 15

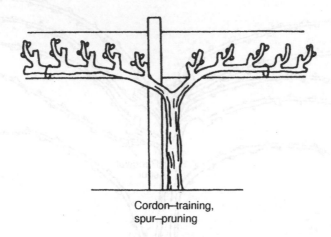

Cordon–training,
spur–pruning

crop. Consequently 85 to 90 percent of the shoot (cane) growth made in the previous season is removed in each year's winter season when the vine is dormant.

Fruiting Wood

The fruiting wood on the vine could be either spurs, with 1 to 4 buds, or canes with 8 to 12 buds. Canes are suitable for varieties which bear small clusters or for varieties which have basal unproductive buds like Thompson Seedless. For the majority of varieties, spur pruning will be satisfactory.

Spur Pruning

On the basal part of the canes, retain two-bud fruiting wood called 'spurs.' You may retain up to 4 buds depending upon the variety. On either side of the arm (trunk), retain 8 to 10, two-bud spurs. The total number of buds retained on the vine should be 35 to 40 depending on the variety, spacing and vine vigor.

Cane Pruning

Retain canes with 8-10 buds as fruiting wood. The number of canes retained should be 3 or 4, keeping the total number of buds below about 35 (could be higher for certain varieties). The final bud number retained will vary with the variety, spacing and vine vigor.

Selection of Buds

Since the vine produces many buds on canes that are needed for next year's crop, the homeowner has to decide on which canes (or arms if vine is old) two-bud spurs should be retained. The buds on canes which are shaded by the leaves of other branches are less fruitful, i.e., they do not bear as many clusters on the shoots compared to those buds which develop from sun exposed canes. Therefore select buds on canes which were sun-exposed during the last season. If two canes are growing very close to each other, retain the stronger of the two and prune the other. The total number of buds retained on the vine depends upon the variety. With certain varieties more buds could be allowed to remain than others because their buds are not very fruitful. The total number of buds per vine should be between 25 and 45 on 12-18 spurs. Leaving too many buds will produce a big crop and weaken the vine in the long run.

By the end of the third growing season, vines are considered mature and are capable of giving a full crop. Prune the vines every year in dormant condition as discussed above. In colder climates, delay pruning till the danger of very cold weather has passed to minimize the damage to the buds.

Disposal of Prunings

Bury or burn the prunings especially if they are infested with insects or diseases.

SOIL FERTILITY AND WATER MANAGEMENT

Grapes are deep rooted. Roots of the vines can grow to depths of 6 feet or more, though most feeder roots (which supply water and nutrients) develop in the top 2.5 to 3.0 feet. The soil at the site should be at least 2.5 to 4 feet deep for proper root development.

Grapevines on shallow soils will develop a shallow root system and are subject to winter root damage or summer drought. Before you plant the vines, dig a few holes 3 feet deep in different areas of your land to determine the soil depth, texture and depth of water table. If the soil is not deep enough, select another location.

Soil fertility is less important than soil depth. Fertility can always be gradually built by the application of organic matter and fertilizers. Well decomposed compost if cheaply available should be incorporated to improve the physical condition of the soil, at the rate of half, 20 lb bag per hole. If the soil is acid as determined by soil analysis and requires liming to correct the acidity, lime should be applied. Consult your cooperative extension agent before applying lime. Very fertile soils promote too much vegetative growth and delay maturity. Gravelly or loamy soils are considered best for grapes. Avoid extreme soil types like sand or heavy clays as grapes grow poorly on them. However, grapes can grow well on a variety of soil types of medium fertility and slightly acid pH. There are commercial soil testing laboratories in most states to test your soil. Contact your cooperative extension agent for information on soil testing.

The soil should have good water holding capacity but be well drained since grape roots are very sensitive to poor drainage. Sandy soils have less water holding capacity compared to loamy soils. If water does not seep through the soil after irrigation, you have poor drainage.

Water Supply

It is important to water thoroughly in times of drought especially during the early years to keep the vines growing well. In most grape growing regions, grapes will grow poorly unless they are irrigated. When the weather gets hot, grapes need irrigation once a week. In the eastern and midwestern United States, due to very high humidity this may not be needed. If you are growing only a few grapevines, use city water but this becomes expensive in the long run. If water is supplied from a well, drip irrigation will be more economical. Sprinklers could also be installed depending upon the size of the planting.

Water high in salts such as sodium and chlorides is toxic to the vines. Have a water sample analyzed if you suspect it has a high salt content.

DISEASE AND PEST CONTROL

There are many insect pests and diseases which can cause failure of crop. The kind of pests depends upon the types of grapes and the regions in which they are grown. Periodically examine your vines for evidence of the pests and diseases. Follow the home garden control measures which are listed in the cooperative extension bulletins of your state.

HARVESTING

Unlike some other fruits, grapes do not ripen if they are harvested in an immature condition. Therefore they should not be picked until fully ripe. The correct stage of maturity will depend upon the variety. The attainment of full color and characteristic flavor and softening of berries are good indicators of maturity. The crop is harvested by breaking or cutting the cluster stems, called peduncles, from the canes. For those varieties which have a very strong, woody cluster stem, use the harvesting shears or a sharp knife with a curved blade to separate the cluster from the shoot. Always hold the clusters by the stem during harvesting. Holding the fruit removes the bloom from the berries and makes them unattractive.

STORAGE

Remove defective berries after harvest and before placing the cluster in the picking container. Place the clusters in a wooden, plastic or styrofoam box or in any picking basket in a single layer. If the fruit will not be used immediately, cover the box with plastic into which several holes have been made and place the box in a refrigerator or in any cool room. Keeping the grapes in air-tight plastic bags encourages the development of fungal diseases. If kept in the refrigerator, grapes could be kept in good condition for up to ten days or longer.

TERMINOLOGY USED IN GROWING, TRAINING AND PRUNING GRAPES

Apical: At the end, tip or outermost part.

Arm: The main branch of a trunk on which shoots (canes), flowers and fruits are borne.

Basal: At the base; near the point of attachment of an organ.

Bloom: Refers to flowering. The term also refers to the delicate waxy or powdery substance on the surface of the berry.

Bud: The growing point in the axil of the leaves. An undeveloped shoot usually protected by scales; from the buds shoots arise on which leaves and fruit are borne. The grape bud is a compound bud consisting of 3 growing points; the primary bud being the largest of these growing points, the secondary and tertiary being the two smaller growing points.

Cane: A mature woody shoot which has hardened at the end of the growing season, having several buds.

Clone: A group of genetically identical individuals, all having been vegetatively propagated and derived from a single individual plant.

Cordon: The long arm or arms of a vine usually trained along a wire. Also a method of training in which the vine trunk is allowed to grow to a certain height and then divided to make a bilateral cordon with two arms, one extending in each direction.

Dormant: That time of the year when vines are inactive.

Head: The top of the trunk.

Internode: The portion of the cane between nodes.

Larva: Immature stages between an egg and pupa in insects.

Lateral: A branch of a shoot.

Node: The thickened part on the cane where buds are located.

Pruning: Removing the vegetative parts of the vine to regulate the growth and to train it to keep a certain shape. This is normally done when the vine is dormant.

Shoot: Green growth arising from a bud. The shoot when fully developed bears leaves, flowers, fruits and tendrils.

Slipskin: The pulp slides easily out of skin of the berry.

Spur: A cane pruned to 4 or fewer buds; it is a fruiting unit.

Sucker: A shoot from a bud below ground.

Tendril: A slender, curved structure borne at some of the nodes on the shoot which provides support to the shoot.

Trunk: The above ground main and permanent stem or body of the vine between the roots and the place where the trunk divides to form branches.

Water sprout: A shoot from a bud on the trunk or arms.

NOTE

Most states in the United States, and some provinces in Canada, have published extension bulletins on grapes. Only a few are listed here. Contact your cooperative extension office for a list of grape publications.

SUGGESTED READINGS

Ahmedullah, M. and D. Mair. 1979. Symptoms of Grape Disorders in Washington. Extension Bulletin 0722. Cooperative Extension, Washington State University.

Ahmedullah, M. 1980. Grape Varieties, Clones and Rootstocks for Washington and Oregon. Oregon Hort. Soc. Proc. 215-222.

Ahmedullah, M., and D. G. Himelrick. 1990. Grape Management. Chapter 10 in *Small Fruit Crop Management*, Eds. G. J. Galletta and D. Himelrick. Prentice Hall.

Andris, H. and P. Elam-Wenzel. 1985. Growing Quality Table Grapes in the Home Garden. University of California, Cooperative Extension Service.

Hardin, C. N., and S. H. Blizzard. 1981. Growing Grapes in West Virginia. Circular No. 117 (Revised). West Virginia University, Agricultural and Forestry Experiment Station.

Hegwood, C. P., R. H. Mullenax, R. A. Haygood, T. S. Brooks and J. L. Peeples. 1993. Establishment and Maintenance of Muscadine Vineyards. Bulletin 913, Mississippi Agricultural & Forestry Experiment Station, Mississippi State University, Mississippi.

Howell, G. S. 1986. Grape Varieties for Michigan's Vineyards. Cooperative Extension Service, Michigan State University. Extension Bulletin E-1899.

Mortensen, J. A. and C. P. Andrews. 1981. Grape Cultivar Trials and Recommended Cultivars for Florida Viticulture. Proc. Fla. State Hort. Soc. 94: 328-331.

Reisch, B. I. and R. M. Pool. 1993. A Catalogue of New and Noteworthy Fruits. New York State Fruit Testing Association, P. O. Box 462, Geneva, NY.

Stang, E. J., D. L. Mahr and G. L. Wart. 1985. Growing Grapes in Wisconsin. Bulletin A 1656. University of Wisconsin, Cooperative Extension Service, Madison, Wisconsin.

Strik, Bernadine, C. 1993. Grape Cultivars for Your Home Garden. EC 1309, Oregon State University Extension Service.

Raspberries

Marvin P. Pritts

Canes that are 30 feet long–fruits that are red, yellow, orange, purple, and black–plants that fruit in the summer and others that fruit in the fall–stems that are thorny and square or round and smooth–from the Arctic Circle to near the equator. This describes some of the diversity that exists in more than 200 species of plants that collectively are called raspberries.

Raspberries belong to the genus *Rubus* that also contains black-berries. They differ from blackberries in that the receptacle (core) stays attached to the plant when raspberries are picked, leaving a hole in the bottom of the fruit. Otherwise, the plants are similar, sharing thorny canes and producing an aggregate fruit consisting of many tiny drupelets. Raspberries belong to the Rose Family, but they don't look like garden roses. The number of species in this family is very large, containing many other horticulturally impor-tant plants such as apples, cherries, peaches and strawberries.

Despite the diversity that exists among raspberries, the cultivated plants are derived almost exclusively from just two species. *Rubus idaeus* and *R. occidentalis* are the wild red and black raspberry, respectively, and from them have been developed cultivated red

Marvin P. Pritts is Associate Professor and Specialist in berry crop production, Cornell University, Ithaca, NY.

Illustrations are by Marcia Eames-Sheavly, Extension Specialist, Department of Fruit and Vegetable Science, Cornell University.

[Haworth co-indexing entry note]: "Raspberries." Pritts, Marvin P. Co-published simultaneously in *Journal of Small Fruit & Viticulture* (Food Products Press, an imprint of The Haworth Press, Inc.) Vol. 4, No. 3/4, 1996, pp. 189-225; and: *Small Fruits in the Home Garden* (ed: Robert E. Gough, and E. Barclay Poling) Food Products Press, an imprint of The Haworth Press, Inc., 1996, pp. 189-225. Single or multiple copies of this article are available from The Haworth Document Delivery Service [1-800-342-9678, 9:00 a.m. - 5:00 p.m. (EST). E-mail address: getinfo@haworth.com].

(Photo 1), black (Photo 2), yellow (Photo 3) and purple (Photo 4) raspberries. The purple type is a cross between the black and red raspberry. Yellow raspberries are produced by mutant red raspberry plants. Cultivated raspberries have improved yields and fruit size compared to their wild relatives.

Most gardeners can get good crops of about 60-70 pints of fruit from 100 feet of row in a well-managed planting, although 100 pints is not an unreasonable goal. Of course, some varieties are more productive than others. With the proper selection of varieties, you can harvest raspberries of different colors from early summer

PHOTO 1. Red raspberries.

PHOTO 2. Black raspberries.

N.Y. 628

through fall. The expense of supermarket raspberries is an incentive to grow your own–the taste of home-grown raspberries is an even greater reward.

CLASSIFICATION AND ORIGIN

Wild raspberries have been harvested for thousands of years. The first recorded raspberry harvest was from Mt. Ida in Greece in AD 45, so the red raspberry was named "idaeus" (meaning "of Ida") to reflect this. Interestingly, modern botanists have failed to find raspberries growing there, and some have suggested that the Ide Mts. in Turkey were the actual place of origin. Nonetheless, raspberries were cultivated in Roman gardens by the fourth century and widely used for medicinal purposes throughout Europe. In the late 1700s, four varieties were available in the United States; two of these were wild selections from the United States, and two were from England

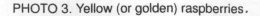

PHOTO 3. Yellow (or golden) raspberries.

where the plant also grows wild. By 1829, 23 cultivated varieties were known in England, and 20 varieties were sold by Prince's Nursery in New York in 1832.

Interest in raspberries continued to increase, and the New York State Experiment Station began a breeding program for this fruit in the late 1800s. By 1925, 415 varieties were available and many thousands of acres of raspberries were grown throughout the northeastern United States. Much of the black raspberry fruit was dried, or used to make food dyes. As several other state experimental stations began to breed raspberries, disaster struck the fruit growers.

PHOTO 4. Purple raspberries.

Raspberry fields began to turn yellow and die from a mysterious disease (Photo 5). Scientists discovered that the disease was caused by a virus that was spread by aphids sucking juices from infected plants and reinfecting healthy plants. This group of viruses, called raspberry mosaic, is systemic, so daughter plants propagated from infected mother plants also were infected. It was, therefore, difficult to reestablish a healthy field. This disease was a major cause for the decline of raspberry production in the Northeast, and its rise along the West Coast where the disease was not found. The West's cooler summers and milder winters also were more favorable for raspberry production than conditions in the East.

New propagation methods and virus detection techniques have allowed growers to plant healthy fields of raspberries. This, coupled with a high demand for the fruit, have encouraged growers to establish raspberry plantings throughout the country, even in the eastern United States. Because raspberries are tasty but very perishable, they command a high price in the store. This high demand and price

PHOTO 5. Mosaic virus disease in black raspberries.

continue to motivate growers to plant raspberries, despite the fact that they are perhaps the most difficult crop to grow and deliver to market in good condition.

The three major commercial raspberry producing regions today include Russia, Europe (mostly in Poland, Hungary, Germany, and the United Kingdom), and the Pacific Coast of North America (British Columbia, Washington, and Oregon). Much of the fruit produced in these regions is harvested mechanically and processed. In other production regions, such as eastern North America, nearly all the production is for the fresh market. But you don't have to rely on commercial markets. Grow your own. They are pretty plants that are very responsive to the care that you give them, and they bear plenty of luscious, nutritious fruit.

HUMAN NUTRITION

The raspberry fruit is mostly water (about 87%). A typical ripe raspberry fruit also contains 5 to 6% sugar. The amount of acid in the fruit increases early in fruit development, then, as in most other fruit, decreases during ripening. The balance between the sugars and acids is important for raspberry flavor. A fruit with low sugar and high acid will taste tart; one with high sugar and low acid will taste bland. Typical pH of a ripe raspberry fruit is 3.0-3.5 and the ratio of sugars to acids (%w/w) is about 1.0. Fruits grown under warm, dry summers are sweeter, less acid, more aromatic, and more highly colored. Hot weather reduces the aroma of the fruit, and wet weather reduces the sugar content.

Raspberries contain ellagic acid which has been shown to prevent certain cancers in laboratory animals. Raspberries contain more of this substance than any other fruit that has been tested. Scientists are now trying to determine how much ellagic acid will give protection against these cancers.

Raspberry fruit contains small amounts of vitamins; only vitamin C is present at a significant level. However, raspberries are a rich source of soluble fibers. These have been shown to help prevent heart disease by lowering abnormally high levels of blood cholesterol. Soluble fibers also help diabetics by slowing the release of carbohydrates into the blood stream and maintaining a more even blood glucose level. There is no Recommended Dietary Allowance (RDA) for fiber, but the National Cancer Institute has set a goal of 20-30 g of fiber per day for adults, the amount contained in 1/2 to 1 pound of raspberry fruit.

PLANT CHARACTERISTICS

The raspberry growth habitat is unusual. The underground root system and crown are perennial. Canes grow from underground buds in spring, achieving a height of several feet by autumn. These first year canes, called primocanes, produce buds in the axils of leaves in late summer and autumn. After winter, canes entering their second year are called floricanes. Axillary buds (those produced at the point of attachment of last year's leaves) grow from floricanes in spring, producing leaves, flowers, and fruit. After fruiting has occurred, the floricanes begin to age and eventually die. New primocanes are produced each spring, so after the second year, fruit is produced every year (Figure 1).

Some raspberry varieties are capable of producing flowers at the tips of primocanes if the growing season is sufficiently long (Photo 6). With primocane-fruiting varieties, fruit ripens in late summer or early fall, and canes do not senesce right away. If these canes overwinter, additional fruit is produced on axillary buds the following summer (Figure 2). These varieties are occasionally called "everbearing," "double-cropping," or "fall-bearing" because you can harvest two crops each year. Commercial varieties produce red or yellow fruit.

FIGURE 1. Life cycle of a summer-bearing raspberry cane.

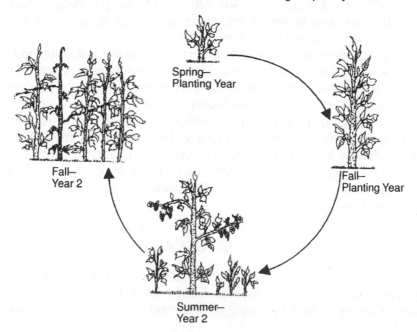

Spring—
Planting Year

Fall—
Year 2

Fall—
Planting Year

Summer—
Year 2

PHOTO 6. A red primocane-fruiting variety.

ADAPTATION

Cultivated raspberries have somewhat exacting climatic require-ments, even though wild raspberries grow under widely varying conditions. An important consideration is temperature. Raspberries are intolerant of high temperatures during the growing season. Pho-tosynthetic rates fall dramatically when temperatures exceed 80°F. Optimal growth occurs between 60 and 75°F. Raspberries also have a chilling requirement of about 800 hours (below 45°F), although this is quite variable among varieties. In areas where temperatures fluctuate dramatically, as in the Midwest, cold temperature injury is

FIGURE 2. Life cycle of a fall-bearing raspberry cane.

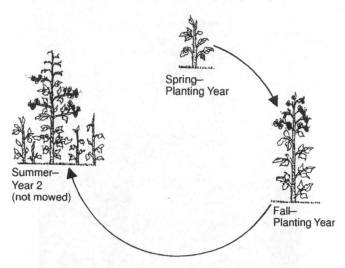

common when warm, early spring temperatures are followed by subfreezing temperatures.

Minimum temperatures in winter also affect the survival of floricanes into the fruiting year. In general, red raspberries are hardier than black raspberries, although major differences exist among varieties. If a red raspberry variety originates from eastern Canada or the northeastern United States, it will likely tolerate temperatures to −20°F in mid-January. If the variety originates from western North America, Scotland or England, then injury may occur below −5°F. Plants can be injured at warmer temperatures than these if unusually cold weather occurs earlier or later than midwinter.

The best places to grow raspberries are those where temperatures are not too cold in the winter, not too hot in the summer, and where spring fluctuations are minimal. In North America, this area is in western Oregon, Washington and British Columbia. Red raspberries are not well adapted to southern parts of North America; however, a few varieties have been bred for heat tolerance. Certain cultural practices, like the use of rowcovers on primocane-fruiting types in spring, will accelerate growth, flowering and fruiting in climates where the growing season is too short or cold for raspberries.

The amount of sunlight during the growing season is directly related to growth and productivity. Areas that receive lots of sunshine during the growing season can produce good crops of raspberries, if temperatures are right and enough water is available from the prebloom through ripening stage.

Rain is the bane of raspberry growers. Because raspberries are susceptible to many fungal diseases, heavy or frequent rain during flowering or fruiting can facilitate disease outbreaks and moldy berries. On the other hand, adequate water is necessary for optimal growth. It is best to supply water by drip irrigation rather than through natural rainfall so the leaves and flowers remain dry. Again, the Pacific Northwest receives most of its rain during the winter, and is generally sunny and dry during spring and summer, making it a good region for raspberry production.

VARIETY SELECTION

More than 40 summer-bearing red raspberries are grown commercially, and these change with the release of new, improved varieties. Among the major varieties originating in North America are Boyne, Canby, Chilcotin, Killarney, Reveille, Skeena, Taylor, Titan, and Willamette. The Glen and Malling series are important varieties from the United Kingdom. In southern states, Dormanred, Southland, and Bababerry are grown.

The major variety of fall-bearing raspberry is Heritage; other important varieties are Amity and Autumn Bliss. Fall-bearing varieties are popular because pruning is simpler than with summer-bearers, and they have fewer pest problems. Yellow varieties are available, but are not widely grown because they tend to have soft fruit. However, their flavor is excellent. Recently, yellow mutations of Heritage have been named and propagated. These include Goldie and Kiwigold. They have many of the characteristics of Heritage such as high productivity and firm fruit.

At least 13 varieties of black raspberries are grown in North America. The most successful are Allen, Bristol, and Jewel. Black raspberries are widely grown in western Pennsylvania, West Virginia and southern Ohio. They are not as hardy as red raspberries, and there is little difference among varieties.

Hybrids between red and black raspberries produce purplish colored fruit–some mistake the ripe purple raspberry for an overripe red raspberry. Brandywine is one of the oldest purple raspberries. The fruit is tart but it makes excellent jelly. Royalty is the most popular purple raspberry. Purple hybrids are extremely vigorous and produce large quantities of big, sweet, flavorful berries. Aphids will not feed on Royalty, so this variety does not succumb to mosaic virus disease.

The type of raspberry to grow depends on your climate. Growers in colder climates (Rocky Mountain States, Dakotas, Upper Midwest, New York, New England, and Canada) should plant primarily red or purple raspberries, as black raspberries don't have adequate hardiness to come through winters without some cane dieback. In warmer areas (Mid-Atlantic, Lower Midwest, Lower West Coast), red raspberries don't grow as well, but black raspberries thrive. Fall bearing types are not sensitive to cold temperatures if they are mowed to the ground every spring, but in colder regions (Canada, Upper New England, Upper Midwest) the growing season may not be long enough to ripen the fall crop. Use Table 1 to select varieties that are adapted to your particular region.

Variety selection is important in determining the length of the harvest season. By selecting an early, mid-season, late-ripening, and a fall-fruiting type, you can harvest berries continuously from the end of strawberry season to the first autumn frost.

Raspberries are mostly self-fruitful, but you will have better crops if you plant more than one variety. Raspberries produce profuse amounts of nectar which attract many species of pollinating bees. Pollination also can occur through wind, so inadequate pollination is rarely a problem. Poor pollination is usually the result of insect damage, cold temperature or a nutritional imbalance.

Select varieties based on their use (processing, freezing, fresh), hardiness, productivity, relative disease susceptibility, fruit quality, and time of ripening. Generally, raspberries are selected as to which have the fewest faults, rather than which have the greatest assets. Although many varieties are available, none are without at least one major fault.

TABLE 1

Variety	Type	Area of adaptation	Outstanding characteristics
Red raspberries			
Autumn Bliss	PF	Wide	Fruit quality Early fall harvest
Bababerry	FF	South	High temperature tolerance
Canby	FF	Pacific Northwest and Mid-Atlantic	Fruit quality and size Early, moderately hardy
Chilcotin	FF	Pacific Northwest	Fruit quality for fresh or processing High yields Not hardy
Dormanred	FF	South	High temperature tolerance
Heritage	PF	Wide	Fruit quality, yield Late fall harvest
Killarney	FF	North	Fruit quality Earliness Cold hardiness
Reveille	FF	Wide	Earliness Cold hardiness
Taylor	FF	Eastern US	Excellent flavor Late summer crop Disease susceptible
Titan	FF	Eastern US	Large fruit size High yields Disease susceptible
Purple			
Royalty	FF	Eastern US	Large fruit size High yields Flavor Disease resistance

TABLE 1 (continued)

Variety	Type	Area of adaptation	Outstanding characteristics
Black			
Allen	FF	Wide	All are similar regarding
Bristol	FF		fruit quality, yield and
Haut	FF		disease resistance
Jewel	FF		
Yellow			
Amber	FF	Wide	Most of these are new
Fallgold	PF		varieties, or they have
Golden Harvest	PF		been recently virus-
Goldie	PF		indexed. Performance is
Honey Queen	FF		still being evaluated.
Kiwigold	PF		

(PF—primocane fruiting; FF—floricane fruiting)

SOIL AND SITE REQUIREMENTS

Raspberry plants require full exposure to sunlight, so don't plant them near trees or buildings. A site that is elevated above the surrounding area can reduce low temperature injury of raspberry canes and provide a measure of protection against late spring frosts.

Poor air circulation causes high humidity around the canes, favoring the development of many diseases such as spur blight, anthracnose, powdery mildew, and the fruit rots. While good air movement is desirable, excessive wind exposure can be a problem. Cold, drying winds can increase the incidence of winter injury. Planting raspberries near a wind break, but not in the shade, will help prevent this kind of injury.

Raspberry plants will not tolerate poorly drained soils. Even temporary water-saturated soil conditions can cause serious injury, including poor cane growth, increased incidence of soil-borne diseases, and plant death. Actively growing raspberry roots will begin to suffocate and die if submerged in water for 24 hours or more. If soil drainage is poor, plant raspberries on raised beds (Photo 7).

PHOTO 7. A new raspberry planting established on a raised bed.

Raspberries grow best on deep, well-drained loams with good water holding capacities, which are well supplied with organic matter. Sandy loams can be successfully used for raspberry production, but raspberries are more difficult to grow on clay soils. On favorable soils, the roots can grow as deep as four feet.

As few as five rainless days in summer might deplete the available soil moisture and stress the plants, so you will need irrigation to obtain consistently high yields. The water should contain little to no salt. Drip irrigation is better than overhead irrigation because it prevents wetting of the foliage, flowers and fruit. Between one and two inches of water per week are required during mid-summer, or 35 gallons per 100 ft of row per day.

Diseases can build up in the soil when certain crops are grown over a period of years. Don't plant black raspberries where tomatoes, potatoes, eggplants or strawberries have been grown during the previous 4 or 5 years, as such sites are frequently infected with *Verticillium* wilt. Broadleaf weeds, especially lambsquarters, redroot pigweed, and those of the nightshade family, also appear to increase *Verticillium* disease. Similarly, if turfgrass preceded the planting, then chafer or Japanese beetle grubs may be present.

Virus diseases are transmitted to raspberries by aphids, and are carried from wild brambles to cultivated crops. To minimize the chance of infection by the many virus diseases, set new plantings as far as possible from wild brambles.

PLANT SELECTION

A number of sources exist for obtaining planting stock, but quality can vary widely among nurseries. Since the cost of plants is minimal relative to the time and energy invested in growing them, purchase the highest quality planting stock available. Poor plant material guarantees a poor planting.

Some people obtain their planting stock from neighbors who offer their raspberry suckers for transplanting. This is not a good idea. Raspberry viruses are systemic (throughout the whole plant), so if the mother plant is infected, the daughter plants also will be infected. Since it is not possible to know with certainty from a visual inspection if a plant is infected with a virus (several years may elapse before viruses cause symptoms), obtaining certified plants from a nursery is recommended.

Nurseries may give you a choice between conventionally-propagated plants or tissue-culture-propagated plants (Figure 3). Dormant suckers or "handles" are canes with one season of growth that were dug after becoming dormant in the fall, and stored as bare roots until shipping. This is the conventional way to purchase red raspberry plants.

Tip-layered canes are the traditional type of black raspberry transplant. The growing tips of the mother plants are covered with

FIGURE 3. Various types of raspberry plants are sold.

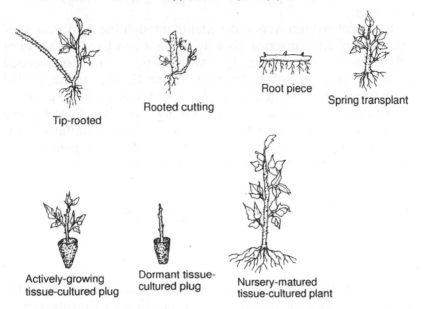

Tip-rooted

Rooted cutting

Root piece

Spring transplant

Actively-growing
tissue-cultured plug

Dormant tissue-
cultured plug

Nursery-matured
tissue-cultured plant

soil in late summer, causing them to root. These are separated from the cane following dormancy and stored as bare roots until transplanting.

Tissue-cultured plants originate in a test tube. Growing tips of a known variety are removed from a virus-indexed source under sterile, laboratory conditions, and placed on a specialized culture media in growth chambers. This small cluster of cells or shoot meristems receive treatments which cause small plantlets to develop. The plantlets are placed in sterile rooting media and grown out in greenhouses. This method, although more costly, provides greater assurance of disease-free plant material than traditional propagation techniques.

Tissue-cultured plants can be either actively growing, dormant or nursery-matured. The latter have been grown in the field for several months prior to digging. Actively growing plants are usually only a few inches tall and must be planted after the danger of frost has passed, but not too late in the growing season. Although nursery-matured plants are larger than actively growing plants from the greenhouse, the latter generally exceed the growth of the former by year's end.

Most raspberry growers order plants through the mail from specialized raspberry nurseries. Usually plants are ordered in the fall or early winter for spring planting. Fall planting is not recommended for colder areas of North America. In the South, mature, potted raspberries can be planted in late fall, but actively-growing tissue-cultured plugs should be planted in spring. Specify a shipping date based upon your projected date of planting. The agriculture department of the state in which the nursery resides is responsible for inspecting the plant material before it is sent to you.

CULTURAL PRACTICES

The Planting Year

You must eliminate perennial weeds, add soil nutrients, adjust the soil pH, and add organic matter before planting your raspberries. If these steps are done the year before planting, then dormant raspberry plants can be set as soon as the soil can be worked in the spring. Plant actively-growing plants after the danger of frost has passed. Fall-planted raspberries should be mulched after planting to prevent rapid changes in soil temperature and moisture. Fall planting of potted or mature, bare-rooted stock is recommended in the South, but is risky in northern areas where temperatures can drop rapidly.

Dormant sucker plants that are purchased for spring planting are usually shipped bare-rooted, and are likely to be dry upon arrival. Soak their roots in water for several hours prior to transplanting. Set them at the same depth as they were before digging; roots should be spread laterally from the sucker stem. Retain about 5 inches of the transplant's stem or handle and prune off the rest. Water immediately after planting. Dormant tissue-cultured plants are handled similarly, except that their roots are not soaked prior to transplanting.

Dormant black raspberry transplants must be oriented so that tips of crown buds are pointing toward the soil surface. The center of the crown should be 3 inches below the soil surface, so that tips of dormant crown buds are approximately 2 1/2 inches deep. Spread roots laterally and slightly downward. Pack soil firmly around the roots to eliminate air spaces, be careful not to damage buds while tamping the soil, and water. Remove the old piece of the parent cane at ground level after plants start to grow.

Succulent red raspberry primocane suckers may be transplanted in early spring (i.e., when suckers are 5 to 8 inches tall) or in fall in the South. They are sensitive to high winds and drought, so irrigation and wind protection may be necessary until they become well established.

Succulent, tissue-cultured plants which have never been exposed to the outdoors require special care. These greenhouse grown plants have shallow root systems which make them sensitive to dry soil and herbicides. They are also susceptible to frost damage and wind. Therefore, plants should be mulched and irrigated, and not planted until the danger of frost has passed. Succulent plug plants are easy to transplant, and growth is more uniform and vigorous than traditional plant material.

Raspberries can be propagated using root pieces collected in early spring. These pieces should be about 6 inches long and contain a bud. Scatter these in a row and cover with about 4 inches of soil.

You can use soluble fertilizer at planting, but dry fertilizer will damage the root system of greenhouse grown plants. Delay dry fertilizer application until several weeks after transplanting. Mulching is the best way to control weeds during establishment, but it should be removed at the end of the year to prevent excessive soil moisture, and to allow new canes to emerge.

Raspberry rows are usually spaced 8 to 12 feet apart to allow one to walk between them when plants are fully grown. Plants are spaced at 2 to 5 feet in the row, depending upon growth habit, vigor of the variety, and training systems which are planned for the crop. Plant black and purple raspberries 3 to 5 feet apart, and red raspberries 2 to 4 feet apart. Red raspberry varieties that sucker profusely may be planted further apart than varieties which form few suckers. Some growers plant primocane fruiting types close together (2 feet) to obtain a high yield earlier in the life of the planting.

The use of raised beds (10 to 12 inches high at peak and 4 to 6 feet wide at the base) may be beneficial in heavy or wet soils to prevent root diseases. Raspberries are sensitive to "wet-feet" and are likely to be short-lived on sites with poor internal or surface drainage.

If you follow preplant soil test recommendations, then nitrogen may be the only nutrient that you will need to supply for the remaining life of the planting. Withhold nitrogen fertilizer from young raspberries until they are growing well, usually 6 to 8 weeks after planting. At this point, supplemental nitrogen in the form of calcium nitrate or high nitrogen fertilizer (15-5-5) is recommended—about 1.5 ounces per plant. Repeat the application in another 6 to 8 weeks. Sprinkle the fertilizer in a ring around the plant. If you prefer organic fertilizers, then there are numerous sources to use. If the soil organic matter content is high, supplemental nitrogen probably is unnecessary the first year.

Most growers apply synthetic fertilizers, even when organic matter is high. This is to ensure adequate amounts during critical periods of growth. Nitrogen is released slowly from soil organic matter and manures, so a sufficient amount may not be available at a critical time if you rely solely on organic sources.

Established Plantings

In older raspberry plantings, nitrogen fertilizer should be applied each spring to promote strong vegetative growth. Fertilization after July 1 can make plants more susceptible to winter injury.

The amount of fertilizer that plants require depends on many factors. Assuming you use ammonium nitrate (34% actual N) as a fertilizer, between 2 and 4 ounces per original plant (or per 3 feet of row) are generally required. This supplies as much nitrogen as 6 to 12 ounces of 10-10-10. Use the lower rate on younger plantings in heavier soils where rainfall is low during the growing season, and use the higher rate in older plantings on sandier soils where rainfall is high or irrigation is used.

Occasionally other nutrients besides nitrogen become limiting, or nutrients that were added before planting leach out of the soil over time. When this happens, growth decreases, leaves may become discolored, and berries can become crumbly. It is often difficult to determine the exact cause of such symptoms. Even if the cause is determined to be nutritionally related, several nutrient deficiencies cause similar symptoms. An easy test to determine the exact nutrient status of an established planting is a leaf analysis. If you are a larger grower, you may want to consider this.

Leaf analysis. Plant tissue analysis measures directly the amount of nutrients in various plant parts. Recommendations are based on the levels of these nutrients at specific times of the year. Unlike visual diagnoses, foliar nutrient analysis can alert the grower when nutrient levels are approaching deficiency so corrective action can be taken. Unlike soil tests, foliar analysis provides accurate results for all essential mineral nutrients.

Currently, recommendations are based on newly expanded leaves collected in early August. Other sampling times or plant parts may prove to be more appropriate for certain nutrients, but until more detailed studies are done, foliar samples collected in mid-summer are the standard.

Collect at least 50 leaves, remove their petioles, and wash them in distilled water. Dry them, place them in a paper bag, and send them to the laboratory for analysis. Samples should be representative of the entire planting. If a particular area of the planting looks poor, sample it separately.

Managing the area between rows. Besides the raspberries themselves, the area between the rows also requires attention. Options for between the rows include continuous cultivation, continuous mulching, mowing the natural vegetation, or seeding a permanent cover crop.

Continuous cultivation is easy, but it destroys soil structure and the field will be muddy after a rain. Continuous mulching can keep the soil too wet, making plants susceptible to root diseases. Natural vegetation may be too competitive with the raspberry plants, and can harbor disease or insect pests.

Generally, the best approach is to seed a perennial grass between the rows, either the autumn before planting, or the autumn of the planting year, and keep it mowed. Perennial grasses tolerate foot traffic, keep out weeds, grow slowly, and do not compete excessively with the raspberry plants. They also help to capture nitrogen that could be lost through leaching, and contribute organic matter to the soil as roots decompose.

TRELLISING AND PRUNING RASPBERRIES

Pruning and trellising have a major impact on growth, fruit number, fruit size, sweetness, and disease susceptibility because of their effect on light interception and leaf wetness.

Primocane-Fruiting Types

Primocane fruiting raspberries produce fruit at the top of first year canes in late summer, and on the lower portion of these same canes in early summer of the second year. Generally, fruit quality of the early summer crop is inferior to both the later crop and the summer crop of floricane fruiting types. In addition, harvesting the early summer crop is difficult because of interference from primocanes, and harvesting the late summer crop is difficult because primocanes are thinner and taller when grown with floricanes. Most growers choose to sacrifice the early summer crop in favor of a smaller, but higher quality late summer crop.

When pruning primocane fruiting types for a single late season crop, canes need only be cut to the ground in winter or early spring. December through February is the ideal time to cut the canes. New canes will emerge and fruit in late summer, and the cycle continues. It is important to cut old canes as close to the ground as possible so buds will break from below the soil surface, rather than from stubs. Canes that are cut from the planting should be removed from the vicinity and destroyed. When canes are removed every spring, winter injury and pest damage are significantly lessened, and pruning is easier. The main disadvantage of fruiting only in fall is smaller overall yields.

In warm climates (USDA Hardiness Zones 6 and 7), the primocane crop can be delayed by mowing the young primocanes a second time when they are approximately 1 foot tall. Pinching the primocanes to stimulate lateral development will also delay fruiting. This is desirable when you want to delay harvest until after the intense heat of July.

Yield in primocane-fruiting types is influenced primarily by the number of canes and the number of berries per lateral. Berry number per lateral is characteristic of a particular variety, but cane numbers can be influenced by the grower. Since there is usually no negative influence of cane numbers on fruit size in the fall crop of primocane fruiting types, you should attempt to maximize the number of canes per area. This can be accomplished by having many narrow rows as opposed to fewer wider rows. Row widths of 18 inches are considered ideal for harvesting.

Primocane fruiting types are top heavy and tend to lean because all of the fruit is produced on the tops of long canes. Most growers find that some type of temporary trellis is necessary during the harvest season to facilitate movement within the aisles. One system that has worked well consists of T-shaped wooden or metal posts approximately 6 feet long with a 2.5 foot crossarm. The posts are pounded into the ground for 2 feet, and the ends of the crossarms contain an eye-screw or other construct which will hold a length of bailing twine. The bailing twine is cheap and disposable, yet strong enough to temporarily hold canes erect.

Floricane-Fruiting Types (Red, Yellow and Purple)

Floricane fruiting types produce fruit from buds on second year canes. Unlike primocane fruiting types, canes must remain intact throughout the winter and until the completion of harvest in summer. During second year flowering and fruiting, first year canes are growing. These primocanes interfere with spraying and harvesting, shade leaves and laterals of fruiting canes, and compete with floricanes for water since each shares a single root system. It is important to minimize this interference.

Typically, primocanes are permitted to grow throughout the season, and the following year, they flower and fruit as new canes emerge. Immediately after fruiting, expended floricanes are cut at ground level and destroyed. In early spring of the following year, the remaining canes are topped to a reasonable height and thinned to a desired number, usually 3-4 canes per ft^2. Remove any winter-injured wood during this process. Select the most vigorous canes to produce the next crop, i.e., those with good height, a large diameter, and no visible symptoms of disease, insect damage, or winter injury (Figure 4).

Since the raspberry plant usually produces excessive primocanes, some of these can be thinned in spring and early summer to allow the remaining ones to grow more vigorously. When some of the

FIGURE 4. Before and after winter pruning of a summer-bearing red raspberry.

Before pruning   After pruning

primocanes are eliminated when they are 6 to 8 inches tall, fruiting canes do not experience as much shading in lower portions of the plant. Harvesting is also easier because fewer primocanes cause less interference. Since there is less demand for water and less shading, fruit size and productivity of lower laterals are increased. An additional advantage of this system is that primocanes are removed at a time when they are small and succulent, as opposed to large and thorny. The major disadvantage is that primocane selection is difficult when leaves are on the plant.

Some growers alternately mow half of the planting each year during the dormant season. Then in the spring after mowing, primocanes will emerge and grow without interference from fruiting canes. The following year, the floricanes will flower and fruit. Although primocanes will also grow in the fruiting year, all canes will be cut to the ground in the subsequent dormant season. Advantages of this system are that no detailed cane thinning or pruning are required, and spray material costs are reduced approximately 50%. Disadvantages include a reduction in fruit quality, berry size, and an overall reduction in yield of approximately 30% for most varieties (since only half the planting is fruiting in any one year).

Special Treatment for Black Raspberries

Black raspberries respond well to primocane pinching in early summer. The density of fruiting buds on a black raspberry lateral is greater than on a main cane; therefore, primocanes can be pinched back when they reach a height of about 28 inches to stimulate branches along the main cane. At least 4 inches of tip should be removed in this operation, and several passes through the field may be required since not all canes grow at the same rate. Ideally, primocanes should be tipped just above a bud to minimize the amount of dead wood that develops between the wound and the bud. Excess dead wood can serve as a site for cane blight infection, especially if tipping precedes wet weather.

At the end of the season, primocanes will be branched with long laterals (Figure 5). The laterals should be supported by the trellis wires by October as wet snow tends to break them off from the main cane. Canes are also more flexible in early autumn so will be less prone to breaking. A significant portion of the laterals may be killed during winter since black raspberries are generally not as hardy as reds. Laterals are shortened in early spring to remove any winter damaged wood, and to maintain berry size. If entire laterals are permitted to fruit, size will be reduced. Most growers shorten laterals to less than 10 inches, although longer laterals can be retained if the wood is thick and healthy.

Trellising Floricane-Fruiting Raspberries

Proper trellising also can help reduce primocane interference and improve production. Without trellising, fruiting canes must be cut rather short in the dormant season to avoid cane breakage or tipping over. Since many of the fruit buds are on the top half of the cane, topping can significantly reduce the productivity of a planting.

Trellising to a single wire 3 or 4 feet above the ground will prevent cane breakage, but allows only a small amount of light to reach the lower portions of canes and forces primocane growth

FIGURE 5. Before and after winter pruning of a pinched black raspberry.

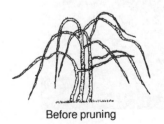

Before pruning

After pruning

toward the aisles. This primocane growth can significantly interfere with spraying and harvesting operations.

Cane interference and competition can be reduced and yields increased by using a trellis which separates the fruiting from vegetative canes. One such system is the V-trellis. Fruiting canes are tied to wires on the outside of the V, and primocanes are permitted to grow in the middle of the V (Photo 8). Spraying, harvesting and pruning operations are made easier since floricanes are pulled to the outside where they are accessible, and primocane interference is minimized. The presence of primocanes in the middle forces lateral growth outward. Yields of several varieties of raspberries have been

PHOTO 8. Red respberries trained to a V-trellis.

increased using a V-trellis, primarily because of the increased amount of light intercepted by the plant canopy.

With any trellis, the end posts must be well anchored. Place trellis posts every 15 to 20 feet in the row. Wires should be loosened before winter to prevent heaving of posts when the metal contracts. Monofilament plastic wire is an alternative to metal, and is much easier to work with.

INSECTS AND DISEASES

Raspberries are susceptible to a number of pests, and few pesticides are available once plants become infested; therefore, you must concentrate on preventing problems before they cause significant yield or plant losses. Good pruning, weed control and water management are the most important practices for preventing pest problems. A few pesticides are available for use on raspberries, but they are effective only if applied at the right time, at the correct concentration, and with the proper equipment. (Note: Regulations regarding pesticide use vary from state to state, and between countries. The Cooperative Extension Service is a good source of information regarding effective methods of pest control.) Following are a few things to watch for.

Weak, Discolored Plants

Insects, mites, nematodes, viruses, fungi and weeds all can be a problem for the raspberry grower. The decline caused by mosaic virus is one example. The tomato ringspot virus is transmitted by nematodes from certain weeds to red raspberries. Dandelion is just one weed that can harbor this virus which causes crumbly berries. Nematodes themselves feed on roots of raspberries, making plants susceptible to root rotting organisms and decreasing their vigor.

Certain varieties of raspberry are susceptible to crown gall. These galls are caused by bacteria which stimulate the plant to form growths on their roots. These galls prevent proper root function, and lead to a general decline of the planting. The bacteria are easily spread from one plant to another.

Nutrient deficiencies and herbicide injury also result in weak growth, and in severe cases, discoloring of leaves. The *Compendium of Raspberry and Blackberry Diseases and Insects* has detailed photographs to help distinguish among these causes of leaf discoloration. Generally, herbicide injury results in distinct patterns of injury in the field, as well as on the affected leaves.

Wilting of Plants

One of the most common diseases of raspberries worldwide is Phytophthora root rot. This disease requires standing water to infect raspberries, so it is most common on poorly drained, heavy soils. Planting raspberries on raised beds will go a long way towards preventing this disease. Plants infected with root rot wilt and die in late spring, even though sufficient water may be present. An examination of the roots will indicate that they are rotted and cannot possibly take up water (Photo 9).

Black raspberries and some red raspberries are susceptible to Verticillium wilt. This disease is most common in young plantings, especially in fields where potatoes, peppers, and tomatoes were grown before. Infected black raspberry canes develop bluish streaks on primocanes, and leaves become yellow and drop. Severely infected plants wilt and die.

Wilting also can be caused by girdling of the canes by borers. Some borer larvae live in the crown and others live in the cane. It is nearly impossible to affect the larvae inside the canes with sprays, and the adult moths are difficult to detect. The best control method is to cut out the infested canes and burn them.

Certain cane diseases can cause wilting of entire canes or of individual laterals. These diseases result in lesions on the canes which are easily visible. As the lesions enlarge, they girdle the canes, and the plants wilt about the girdle. Spur blight forms lesions at the base of laterals, especially on lower portions of the canes (Photo 10). Cane blight lesions can occur anywhere along the cane, especially where wounding has occurred. Anthracnose forms small lesions which coalesce to girdle canes. The latter is more common on black raspberries (Photo 11). All three of these diseases are best managed by improving air circulation through pruning.

Wind can cause wilting of canes, especially when trellises are not

PHOTO 9. Red raspberries infected with Phytophthora root rot.

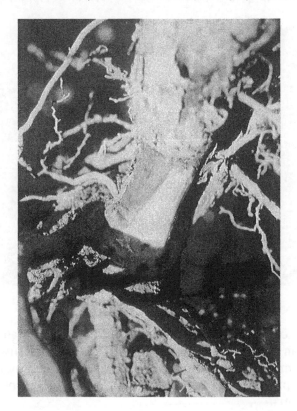

used. The base of the canes become detached, so the upper portion dies. Winter injury damages the water-conducting tissues as well, and wilting may not become evident until hot summer weather occurs. The plant may not be able to keep up with demand for water, so the top of the cane wilts.

There are many causes of wilting in raspberry, so you must determine which is the culprit from among the many possibilities. Trace the wilting back to a point where the plant appears healthy. Look in the area near the transition for damage to the water-conducting tissues. Try to determine what is causing the disruption; this may involve digging plants to examine the root system.

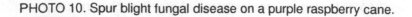

PHOTO 10. Spur blight fungal disease on a purple raspberry cane.

Once the cause of wilting is identified, the appropriate control measures can be implemented. However, prevention is always better than attempting to control the problem. Good pruning and sanitation, and proper water management, are the keys to preventing these problems.

Discolored Leaves

In addition to nutrient deficiencies and herbicides, a few other causes of foliar discoloring occur in raspberry plantings. One of the more common is orange rust (Photo 12). This fungus is visible on the undersides of black raspberry leaves in spring and causes primocanes to be spindly and leaves to cup upwards. The fungus is systemic, so it cannot be cut out. Infected plants should be dug out immediately and destroyed.

Late leaf yellow rust is common on raspberries. This rust does not kill the plant, but it can limit which varieties are grown in certain areas. Late leaf rust causes yellow spots on the upper surface of red raspberries, and sometimes on the fruit of primocane-fruiting

PHOTO 11. Anthracnose fungal disease on a black raspberry cane.

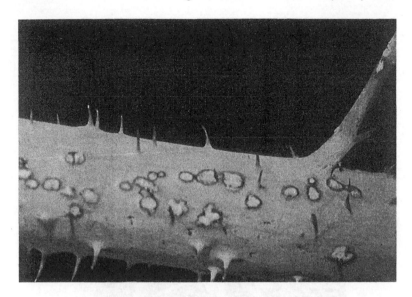

varieties. The undersides of leaves may have small blisters filled with spores. Infected leaves tend to drop prematurely.

During hot, dry summers, mite populations can build up to levels where the upper surfaces of leaves become bronzed. Mites feed on the undersides of leaves, so look there for the tiny mites or the webbing that indicates their presence.

Deformed or Spotted Leaves

Leaf curling can be caused by several organisms, in addition to viruses. One of the more common causes is powdery mildew. This organism primarily infects the undersides of leaves, causing them to curl upwards. The whitish growth of the fungus is usually visible on infected plants.

Leafhoppers feed on leaves, causing upward curling and marginal yellowing on new growth. They can be seen flying among the raspberry plants when the foliage is disturbed. In addition to their direct feeding damage, they can transmit viruses.

Raspberry leaf spot is a fungal disease common on many vari-

PHOTO 12. Orange rust fungal disease on a black raspberry cane.

eties of red raspberry. The disease is characterized by small brown spots on older leaves. In severe cases, leaves become yellow and drop off the plant.

Fruit Malformations

Fruit malformations occur when the individual drupelets are not fertilized properly. A common cause of malformed, crumbly fruit is virus. Usually viruses induce other symptoms on the plant, besides the effects on fruit. When the plants are growing vigorously but the fruit is small and crumbly, the tarnished plant bug may be the culprit. This bug feeds on developing fruit and seeds, injecting a plant toxin as it macerates succulent tissue. The young seeds near the feeding sites do not develop, so the remaining drupelets may be insufficient in number to maintain the integrity of the berry (Photo 13). Both small green nymphs and brownish adults cause this damage. Certain nutrient deficiencies and frost during flowering may result in ovule death and berry deformity.

Several insect species lay eggs on or near fruit, and the developing larvae burrow into the fruit, making it unpalatable. Sap beetles, sometimes called picnic beetles, feed on the ripe fruit and burrow into the berries. Again, fruit infested with sap beetles is unappetiz-

PHOTO 13. Tarnished plant bug on a red raspberry.

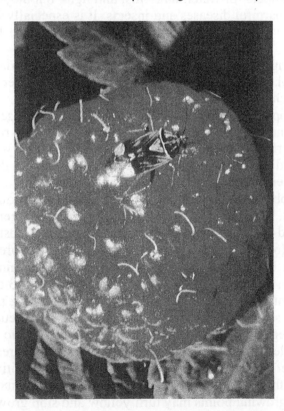

ing. Japanese beetles emerge about the time that the summer crop is ripening, feeding voraciously on leaves and fruit.

Miscellaneous Insect Pests

The strawberry bud weevil can clip off flowers before they open. Obviously, the fruit does not develop. If bud weevils have been a problem in the past, chances are that they will continue to be a problem.

Weed Management

Controlling weeds is essential in a raspberry planting. Not only do weeds compete for water, nutrients, and light, but they can serve as hosts for harmful diseases and insects. It is especially important to control weeds in a young planting when plants are becoming established.

Eliminating weeds prior to planting is the most important step in weed control. The section on site preparation and cover cropping describes techniques to reduce weed pressure.

Plants propagated directly from tissue culturing are sensitive to herbicides and cultivation as they attempt to grow new roots close to the soil surface. Applying a light layer of straw mulch will help to prevent weed seed germination, and reduce the need to disturb the young raspberry root system.

Once raspberry plants become established, they can outcompete many weeds. Weeds that do become established can be removed by hoeing, hand-weeding, or with the careful use of a postemergent herbicide. Most commercial raspberry growers apply a preemergent herbicide in late autumn to suppress weed seed germination the following spring. Typically, a weed-free band between 3 and 5 feet wide is maintained under the row, while the remaining portion of alleyway is seeded to a permanent cover crop of fescues and/or perennial ryegrasses. Postemergent herbicides are used to spot treat certain problem weeds. If you use glyphosate to spot treat, be extremely careful. Raspberry plants are particularly sensitive to this herbicide; although you may not kill the plant if you contact roots or foliage, the growing points may turn yellow and stop growing.

By frequently mowing the area around the planting and between rows, weeds there will not have a chance to flower and set seed. As a result, few seeds will migrate into the raspberry planting. Use caution when applying mulch for weed control in established plantings. Mulch may keep the soil too moist and encourage Phytophthora root rot, or it can be too thick and inhibit primocane growth. If mulch is used, it should be free of weed seeds so as not to introduce additional weeds into the planting.

HARVESTING

Raspberries have one of the highest respiration rates of any fruit. This, coupled with their thin skin and sugary interior, make them very perishable. Many chefs consider fresh raspberries to be the most exotic of all fruits–not because of their rarity, but because of their extreme perishability.

Several steps can be taken to maximize the potential shelf life of a raspberry. For example, harvesting the same planting frequently, at least once every two days, is critical. Fruit harvested just when it is fully ripe will have a much longer shelf life than slightly overripe fruit. The optimum stage of maturity for the raspberry occurs when the berry first becomes completely red, but before any darker hues develop.

Fruit quality usually declines as the season progresses. The earliest berries will likely have the highest quality and largest size for the season. Don't touch berries before harvest; injury caused by picking is often the greatest cause of deterioration.

Harvest and discard overripe or damaged berries because they are susceptible to molds. *Botrytis* is the most common mold of raspberry fruit. Once the mold on overripe berries begins to reproduce, large amounts of inoculum will be present to infect other ripening fruit. Overripe berries also attract ants, wasps and other pests.

Use containers holding approximately 1/2 pint of raspberries for harvest. Wide, shallow containers are preferable to deep containers; raspberry fruit should not be more than 4 layers high. Many different types of containers are available; smaller is usually better.

If the berries are to be stored for later use, you must consider additional factors. Cool berries quickly and then wrap them in plastic to lengthen the shelf life. At 85°F and 30% relative humidity, fruit loses water 35 times faster than at 32°F and 90% relative humidity. Also, for every hour delay in cooling, shelf life is reduced by one day. By harvesting fruit as early in the morning as possible, you can extend shelf life because berries are already somewhat cool.

Refrigerated storage for raspberries can be maintained as low as 30°F. Berries will not freeze at or above this temperature because the sugars in the fruit depress the freezing point.

Fruit that is not used fresh can be frozen for later use. Adding sugar or corn syrup to the berries before freezing will draw moisture out of the cells, causing less damage to the tissue when the ice crystals freeze. Raspberries make excellent juice as well, especially when mixed with a more acidic fruit juice, like cranberry or rhubarb.

In the early 1900s, black raspberry juice was extracted, concentrated, and used as an edible dye for foodstuffs such as meat. Raspberries were also dehydrated for long distance transport. Today, most raspberry fruits are harvested by hand and eaten fresh, or machine harvested and processed into purees, preserves, juice, jam, jelly, dessert topping, pie filling, or yogurt. Raspberry juice makes excellent wine.

USE IN THE LANDSCAPE

Use care when incorporating raspberries into the landscape. Some varieties of red raspberries sucker profusely, spreading beyond the confines of the planted area. Black raspberries do not sucker, however.

Raspberries also require lots of light for optimum fruit production and quality. They can be planted next to woods or walls provided the exposure allows for direct light for much of the day.

Raspberries grow in most garden soils, so they are compatible with most other cultivated plants. They produce fruit within a couple of years after planting, unlike fruit trees or bushes. Their wide adaptability makes raspberries a possibility for landscapes almost everywhere.

Raspberries make effective hedges, forming a thick, thorny boundary. Because raspberries produce fruit throughout much of the year, and because many different colors of fruit are available, raspberries have a definite place in the home landscape.

SUGGESTED READINGS

Crandall, P.C. 1995. *Bramble Production*. The Haworth Press, Inc., Binghamton, NY, 172 p.

Ellis, M.A., R.H. Converse, R.N. Williams, and B. Williamson, eds. 1991. *Compendium of Raspberry and Blackberry Diseases and Insects*. Amer. Phytopathological Society, St. Paul, MN, 100 p.

Himelrick, D. and G. Galletta (eds.). 1990. *Small Fruit Crop Management*. Academic Press, NY, 602 p.

Pritts, M. P. and D. Handley, eds. 1989. *Bramble Production Guide*. NRAES-35, Riley Robb Hall, Cornell University, 189 p.

Strawberries for the Home Garden

E. Barclay Poling

Strawberries are a welcome addition to any home garden. Nearly everyone loves the rich flavor of home-grown strawberries. You do not need elaborate trellises or supports to grow them, and the planting stock is inexpensive compared to that of other small fruits. Strawberries are the first fruit to ripen in the spring, and no other small fruit produces more crop in proportion to the small size of the plant. A garden with as few as 25 plants, occupying less than 100 square feet, can produce 25 quarts or more of bright red berries to eat fresh, to freeze, and to use in making juice, pies, and preserves.

CLASSIFICATION AND ORIGIN

The strawberry is an herbaceous perennial belonging to the Rose family (Rosaceae); which also includes brambles (raspberry and blackberry), peach, and apple. The genus name, *Fragaria*, comes from the Latin *fraga*, referring to the scent of the berry. The modern garden strawberry, *Fragaria anassa* (Fra-*gah*-ree-a an-a-*nas*-a), is derived from 2 native American strawberries, the Virginia "scarlet"

E. Barclay Poling is Professor and Extension Specialist of Small Fruits, Department of Horticultural Science, North Carolina State University, Raleigh, NC 27695-7609.

The section on Everbearing Berries is taken from "Strawberries Like Full Sun–And a Good Deal of Attention," USDA Ag. Info. Bul. 408, pp. 270-271.

[Haworth co-indexing entry note]: "Strawberries for the Home Garden." Poling, Barclay E. Co-published simultaneously in *Journal of Small Fruit & Viticulture* (Food Products Press, an imprint of The Haworth Press, Inc.) Vol. 4, No. 3/4, 1996, pp. 227-257; and: *Small Fruits in the Home Garden* (ed: Robert E. Gough, and E. Barclay Poling) Food Products Press, an imprint of The Haworth Press, Inc., 1996, pp. 227-257. Single or multiple copies of this article are available from The Haworth Document Delivery Service [1-800-342-9678, 9:00 a.m. - 5:00 p.m. (EST). E-mail address: getinfo@haworth.com].

strawberry of eastern North America (*F. virginiana*) and the Chilean strawberry, (*F. chiloensis*), which is found on the Pacific Coast from Alaska to Chile. Early explorers of the New World collected both of these wild strawberries, and chance crosses between the two species in gardens in England and Europe in the middle eighteenth century resulted in a hybrid strawberry, *F. ananassa,* first called the "Pineapple" strawberry by gardeners in Holland. A young Dutch horticulturist, Antoine Nicholas Duchesne, recognized the hybrid character of the modern garden strawberry that arose from the native American species, *F. virginiana* and *F. chiloensis,* and our garden strawberry was named in his honor, *Fragaria* × *ananassa* Duch. The new hybrids combined the size and firmness of the Chilean strawberry with the flavor and productivity of the Virginia strawberry.

There are 11 wild species of strawberry; *F. vesca,* the Wood strawberry, is the most widespread *Fragaria* species in the world. This is the famous French *fraises des bois* native to Europe, but it is also found in Africa, China, North and Central America, as well as in the Andes in South America, and on Pacific Islands. In countries such as France, it is not unusual to see *fraises des bois* being sold in local markets and utilized in pastries and gourmet desserts. The Alpine strawberry, *F. vesca sempervirens,* is a sub-species of *F. vesca* that originated in the mountains of Italy. The Alpines make attractive edging plants, having masses of small white flowers that bear fruits continuously or in flushes, depending upon the growing region.

The modern garden strawberry, *Fragaria* × *ananassa* Duch., is cultivated extensively throughout the world. In fact, the strawberry is economically the leading small fruit grown in the temperate regions of the earth, with large commercial industries located in coastal California, Florida, Mexico, Japan, Poland, Italy, Spain, and Russia.

HUMAN NUTRITION

You can enjoy strawberries in numberless ways, including old-fashioned strawberry shortcake, sundaes, ice cream, pies, jams and jellies, or simply fresh out-of-hand! A single cupful of berries has only 45 calories, comparable to a thin slice of bread or half a cup of whole milk. Nutritionists rate strawberries as an "excellent" source of Vitamin C; one cup of fresh strawberries contains 85 mg of Vitamin C. They are low in sodium and suitable for use in low sodium diets.

PLANT CHARACTERISTICS

A "Compressed Stem"

The strawberry plant has a short thickened stem (called a "crown") which has a growing point at the upper end and which forms roots at its base (Figure 1a). New leaves and flower clusters arise from "fleshy buds" in the crown early in the spring. Under good growing conditions, a side stem, or "branch crown," will develop from additional growing points called "axillary buds." From a cultural viewpoint, it is desirable to have one or more branch crowns form during the summer and early fall, as each of these will add to the yield of the main strawberry crown by producing its own inflorescence, or flower cluster.

Spreading Habit

Strawberries spend much of their energy producing "horizontal stems," (runners, or stolons) to carry their leaves to more favorably-lit places (Figure 1). A single strawberry plant can claim a surprisingly large garden area over just a few years' time, with its runners.

The swollen end of the runner stem (plantlet) develops into a runner plant, with roots on the underside and a shortened stem (crown) and leaves above. The first generation runner plants, in turn, send out more runners in a spreading, step-wise fashion. We usually do not limit the number of runner plants, but we do keep individual plant rows to 14 to 24 inches in width in the "matted row" system of training, explained later.

Long days and warm temperatures promote the production of runners. Buds in the axils of the leaves begin to develop into run-

FIGURE 1. (1a) The strawberry crown is a compressed stem at soil level that gives rise to leaves, runners, roots, flowers, and fruit. (1b) Runner production occurs during the long days and warm temperatures of summer. Then, in the short, cool days of fall, runnering stops and flower buds form within the plant crown.

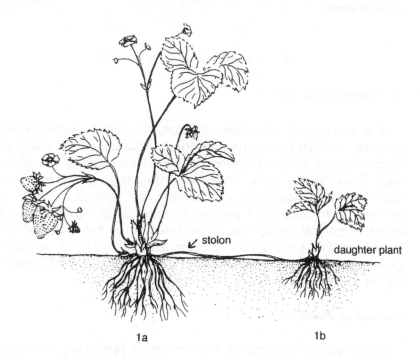

stolon

daughter plant

1a

1b

ners in late spring. This continues through the summer months and often into the fall in lower parts of the South. Commercial strawberries are propagated by runners.

Strawberries Are Not Climbers

Strawberries spread horizontally by runners; they do not "climb" as some nurseries would have you think. They may be trained to climb if you tie the runners to a trellis or a wire cage. The completely "aerial" strawberry plant will not tolerate northern winters.

Roots

The underground parts of the strawberry may be easily ig-
nored—"out of sight and out of mind"—but they do anchor the plant
and capture water and nutrients necessary for photosynthesis. When
the succulent leafy parts of the strawberry die in winter, it is the food
stored in the roots and the dormant stem (crown) that enables this
perennial to quickly regenerate new foliage the following spring.

Strawberries are "shallow-rooted," but in well-drained, sandy
loam and loamy soils, strawberry roots may extend to more than 1
foot deep. There are usually 20 to 35 primary roots and thousands
of small rootlets in a good strawberry root system. Normally, roots
live only a few months, or sometimes up to one year, and they are
continuously replaced by new roots. Formation of new roots mainly
depends upon the temperature, aeration, and moisture in the soil. In
the middle of summer, the formation of roots slows down. Strawberry roots grow well in periods of cooler soil temperatures (spring
and fall), provided there is good soil moisture and the soil environment is free of pests (nematodes, diseases, and insects).

Leaves

The leaves are borne along the crown on petioles (leafstalks)
arranged in spiral fashion around the crown. Strawberries have
compound leaves in which the blade (flattened part of a leaf) is
divided into 3 separate leaflets, called a "trifoliate." The strawberry
leaf captures light, the source of energy used by plants for food
manufacture in photosynthesis.

As days become progressively shorter in late summer and fall,
runnering stops and flower buds form within the plant crown(s).

FLOWERING TYPES

Junebearers

Growth in Junebearer strawberries is affected greatly by temperature and length of the daylight period (photoperiod). Runner pro-

duction occurs during the long days and warm temperatures of summer. Then, in the short, cool days of fall, runnering stops and flower buds form within the plant crown. The flower clusters that develop inside the upper portion of the strawberry crown in the fall emerge in early spring. Berries begin to ripen 4 to 5 weeks after the first flowers open and continue to ripen for about 3 weeks. Toward the end of the harvest period, when days are long and warm, plants again grow runners, which produce new plants. Examples of Junebearer types include such varieties as 'Earliglow,' 'Cardinal,' 'Allstar' and 'Honeoye.'

Everbearers

Everbearing strawberry varieties behave somewhat like "long day" plants. Though everbearing strawberries will flower best under the influence of long days, a short day condition will not necessarily prevent flowers from forming. Everbearing strawberries typically produce crops in spring and late summer/fall. The combined yield of the spring and late summer/fall crops does not generally equal the yield of a single crop of a Junebearer variety. Examples of everbearer types would include the varieties 'Ft. Laramie,' 'Ozark Beauty' and 'Ogalla.'

Dayneutrals

Strawberries that are dayneutral have the distinction of being able to form flower buds during days of any length (thus, dayneutral) and in any season of the year provided temperatures do not exceed 80°F (27°C). 'Tristar' and 'Tribute' are newer dayneutral varieties. The distinction between everbearing- and dayneutral-type strawberries has little practical meaning for our purposes—both types will produce two crops a year. Under the long days of summer, everbearers and dayneutrals can form flowers (if it is cool enough), which produce a second crop of berries in late summer and fall up to first frost.

It is simple to remember that Junebearers, everbearers and dayneutrals will all form flowers in the fall which develop into berries the following spring and early summer.

FLOWER AND FRUIT CHARACTERISTICS

The principal parts of the strawberry flower are shown in Figure 2. Sepals are the small green leaflike structures below the white petals–they enclose the flower at the bud stage, and later on this leaflike tissue is referred to as the berry's calyx, or "cap." The strawberry flower has 5 sepals. The stamens are the "male" parts of the flower that discharge pollen to fertilize the "female" parts of the flower, called the pistils. The numerous pistils are borne on a roundish or conic-shaped flower-supporting stem called the "receptacle." At maturity, the receptacle becomes the enlarged, juicy "berry."

Botanically, the red fruit we call the "berry" is an enlarged flower stem (receptacle) with many seeds imbedded in the surface. Actually, what look like seeds really are the "true fruits," properly referred to as achenes. Inside the dry ovary wall of each achene is a real seed (ovule) with the potential of becoming a strawberry plant. This offspring is not likely to have any of the desirable horticultural characteristics of the parent strawberry. To preserve strawberry varieties which yield superior fruit, strawberry propagation is generally accomplished by taking runner plants that are identical in genetic makeup to the "mother plant." Primary berries are not only the largest and first to ripen; they have the most seeds. Secondary berries ripen next and are the next largest in size. Tertiary berries

FIGURE 2. Principal parts of the strawberry flower: (a) receptacle; (b) pistil and fruit wall; (c) anther; (d) sepal; (e) petal.

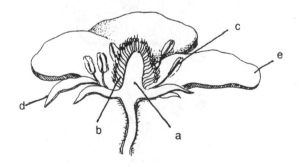

ripen still later and are the third largest. Quaternary berries are the smallest berries; they ripen last (Figure 3).

VARIETY SELECTION

One of the many advantages to having your own garden is the privilege of choosing the varieties you prefer. The strawberry varieties at the supermarket are produced commercially and of necessity are varieties that were not selected so much for their flavor or aroma as for their ability to withstand long shipments and many handlings. The home gardener, on the other hand, can choose varieties on the basis of quality considerations and the exigencies of the climate in which he or she lives. Strawberries are bred and selected especially for specific climatic areas, and few Junebearers, or short day strawberry varieties, are adapted to climates outside the areas in which they have been developed. In the United States, southern varieties require little winter cold to break rest and, as a rule, are not hardy in the northern states and Canada. In contrast, northern strawberry varieties have long chilling requirements and are very slow to flower when planted in the South. Occasionally, a strawberry variety developed for one climatic area is adapted to other regions. The very high quality 'Earliglow' variety is an example of a Junebearer selected for the northern United States that is also well-adapted to

FIGURE 3. Fruit cluster: (a) primary fruit; (b) secondary fruit; (c) tertiary fruit; and (d) quarternary berries.

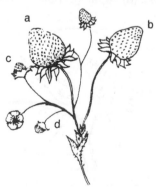

gardens in Virginia, Tennessee and western and upper piedmont sections of North Carolina (it is actually a "Maybearer" in this part of the country).

There are hundreds of varieties available for gardens in the United States and Canada. Tables 1a and 1b give the top regional

TABLE 1a. Short day strawberry varieties recommended by state research and extension specialists in the United States.

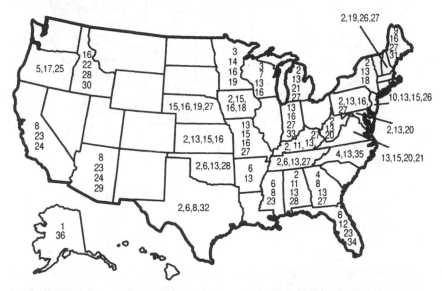

1	Alaska Pioneer	19	Kent
2	Allstar	20	Lateglow
3	Annapolis	21	Lester
4	Apollo	22	Micmac
5	Benton	23	Oso Grande
6	Cardinal	24	Pajaro
7	Cavendish	25	Rainier
8	Chandler	26	Raritan
9	Cornwallis	27	Redchief
10	Darrow	28	Scott
11	Delite	29	Sequoia
12	Douglas	30	Shuksan
13	Earliglow	31	Sparkle
14	Glooscap	32	Sunrise
15	Guardian	33	Surecrop
16	Honeoye	34	Sweet Charlie
17	Hood	35	Titan
18	Jewel	36	Toklat

TABLE 1b. Everbearing and dayneutral strawberry varieties recommended by state research and extension specialists in the United States.

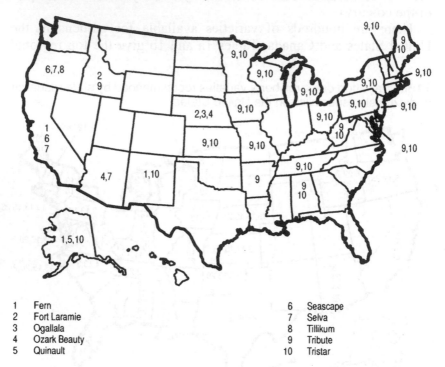

1	Fern	6	Seascape
2	Fort Laramie	7	Selva
3	Ogallala	8	Tillikum
4	Ozark Beauty	9	Tribute
5	Quinault	10	Tristar

picks of some 35 researchers, extension specialists and noted horticultural authorities. The main factors taken into consideration in developing these variety suggestions were (a) state/regional climatic adaptation; (b) yield potential; (c) fresh and processing quality; (d) season of ripening; and, (e) disease resistance. Remember, even the best cultural practices cannot overcome the handicap imposed by poorly adapted varieties. Plant only varieties adapted to your region.

Ripening Seasons

Varieties are listed in nursery catalogs as "Junebearing," "everbearing," or "dayneutral." The Junebearing plants yield only one crop in the spring and/or early summer, but their berries are generally considered to be superior in dessert quality than those of everbearing plants. The dayneutrals 'Tristar' and 'Tribute' are generally considered to be comparable in quality to many Junebearers. If space is not limiting, you may wish to select 2 or 3 different June-bearing varieties with slightly different picking seasons. For a given Junebearing variety, fruit ripens for about 3 to 4 weeks, and under high temperatures it may only last 2 1/2 to 3 weeks. By selecting early, midseason and late season varieties, you can extend the harvest season to up to five weeks in most areas. Remember, each variety comes in a bundle of 25 plants, and 25 plants will occupy 75 to 100 square feet, depending upon plant spacing and training system.

Everbearers and Dayneutrals

Since the everbearing strawberries originated in the northern U.S. and Canada, they succeed best in those areas. They generally do poorly in the southern states because the hot weather stops flower bud development for much of the summer. Everbearers and dayneutrals are recommended only for southern New Jersey west to northern Missouri and northward. However, some gardeners in mountain areas in the south claim success with 'Ozark Beauty.' Typically, a small to medium sized crop is produced by everbearing strawberries in the spring/early summer, and then the plants go into a short "rest" period during which more flower buds are initiated. These buds will begin to blossom in mid-summer, and produce a fairly good crop during August, September and October, or until the first fall frost. The dayneutrals 'Tristar' and 'Tribute' have a fruiting habit similar to the everbearing strawberries. The management of everbearers and dayneutrals will be discussed in a separate section as their culture differs somewhat from the Junebearers.

Ordering

Purchase from a reputable nursery to be sure of getting quality plants true to name and certified to be free of insects, diseases, nematodes and viruses. There are no apparent visual differences between virus-free and ordinary planting stock, and the only way you can be sure of having essentially virus-free plants is to order only certified plants (Table 2). The extra cost is worth the expense when you consider that "virus-free" plants can yield 50 to 75 percent more fruit than plants from ordinary stock. Diseased or low vigor plants can mean a delay in harvest.

Dormant strawberry plants are best for spring planting in the home garden. Order them by late fall/early winter. Dormant plants are typically dug by the nursery in early winter and held in cold storage at a temperature of 29° to 30°F ($-2°$ to $-1°$C) and relative humidity of 85% until planting or shipping time in the spring.

SITE AND SOIL CONSIDERATIONS

Choose the sunniest site for strawberries. Light and warmth are necessary to ripen the fruits and to promote the development of fruit buds in late summer and fall for the next year's crop. If you live in the country where there is plenty of room for wildlife as well, it may be wise to have the strawberry patch near the house under your frequent surveillance, especially during berry ripening.

On level sites, orient your strawberry rows north to south to provide both sides of the rows with equal amounts of sunlight and warmth. If the rows run east to west, plants on the north side of the row will ripen two to three days later than those on the south side. However, the slope of the garden and shape of the strawberry plot are more important considerations than the aspect of the rows.

TABLE 2. Definitions for registered and certified strawberry plants.

Registered plants: These have been grown under State/Province supervision, and the word "registered" on the bundle label indicates the plants are substantially virus-free, the best that can be obtained (these are considerably more expensive, and generally not sold to home gardeners).

Certified plants: Also grown under State/Province supervision. Certification indicates that the plants are free of most noxious diseases and insects; however, they may carry viruses. Certified plants are the best available of some varieties.

Adapted from *Growing Fruits and Nuts*, USDA Info. Bul. 408, pp. 226.

Always plant on a gentle slope protected from strong winds, whenever possible.

A "light" spring frost generally will not harm new strawberry plants, though open blossoms are injured by temperatures of 31°F (−1°C) or below. Frostbitten blossoms have darkened centers (Figure 4). You can protect blossoms from frost by covering plants with 2 to 3 inches of straw, old cloth, paper, or row covers. Plastic sheets give little or no protection. Sprinkling plants with water continuously during the frost period will also give protection, since the change of water to ice on the plants releases heat.

Strawberry plants have a shallow root system and cannot stand severe drought or competition from nearby weeds and trees which deprive the plants of sunlight and moisture. If drought comes during the growing season (transplanting, before or during harvest, and during fall fruit bud formation), irrigate enough to wet the soil 6 to 8 inches deep once a week.

Strawberries are adapted to all well-drained soils. Plants on lighter soils (sandy, sandy loam) warm up earlier in the spring and produce an earlier crop. Light soils drain well, allowing garden work and harvesting sooner after rain than heavier soils (clay, silt). Strawberries can be grown successfully on loam, silt loam, and silty

FIGURE 4. The strawberry blossom on the left is frost injured. The strawberry blossom on the right has not been frost injured.

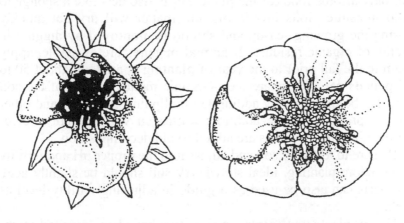

clay loam soils, but yields may be reduced on heavy clay soils because of poor drainage and the resulting poor root development. Soils waterlogged for long periods must be drained, or extensive root death can occur. Where it is impossible or impractical to drain the soil, plant on ridges 6 to 9 inches high and at least 15 inches wide at their base. Ridging attachments are made for certain rototiller models.

A strawberry bed can produce for three to four years, but weeds are the principal cause of a strawberry garden failing in one year or less! Quackgrass (*Elymus repens*) in the north and bermuda grass (*Cynodon dactylon*) in southern growing areas are examples of very fast-spreading perennial grasses that can choke-out newly set strawberry plants, sometimes in less than a month after planting! Vine-like broadleaf weeds such as morning-glories (*Ipomea* sp.) will shade out and even strangle strawberry plants. Other annual broadleaf weeds such as henbit (*Lamium amplexicaule*), chickweeds (*Stellaria* spp.) and Virginia pepperweed (*Lepidium virginicum*) can produce enough seedlings from just a few plants to carpet an entire garden area if allowed to reach the "seeding-out" stage. In the year previous to planting, eradicate the perennial weeds. Use Roundup or similar weedkillers to rid the garden of noxious perennial weeds. Pay particular attention to preventing weeds in or near the garden from reaching the seed stage.

A good supply of organic matter in the soil is important. Humus not only affords food for the plant, but it also acts like a sponge to hold moisture. Soils low in organic matter will dry out quickly during the growing season and should be improved by digging in plenty of organic residues. If animal manures are available, apply them in the fall prior to the year of planting at a rate of about 50 to 75 pounds per 100 square feet. Organic matter content of the soil can also be increased by adding lawn clippings, sawdust and other similar products. Some additional nitrogen will probably be required if organic residues are added to aid decomposition.

Be sure to test your soil and adjust it to the proper pH and fertility level before planting. Ideal strawberry soil should be slightly acid. Soil tests can also be used as a guide in adjusting fertility level as well.

Do not plant strawberries on sites that have been cropped intensively for some time. Often the soil has lost much of its fertility, and the physical properties and tilth of the soil are impaired. Strawberries should not follow any solanaceous crop (e.g., potatoes, tomatoes, peppers, or eggplants) for at least 4 to 5 years. These precautions will help reduce the incidence of serious root diseases such as verticillium wilt and black root rot. Also, wait two years before planting strawberries in ground that is in sod. While this soil may be comparatively high in organic matter, it is likely to be infested with white grubs. These feed on strawberry roots and frequently cause considerable injury and loss of new plants. If it is necessary to plant an area that has been in sod, turn or spade the sod at least one year in advance of planting. Weed problems in plantings set in newly turned sod can be overwhelming. It is always better to work the sod down and keep the area cultivated for at least a year before planting strawberries.

TIME OF PLANTING

Planting by the moon does not seem to be important, but traditional recommendations indicating that dormant strawberry plants should be set "as soon as the ground can be prepared in the spring," can lead to problems in some years. Plants that have been stored long-term at 30°F (− 1°C) at the nurseries will not tolerate post-planting temperatures of 22°F (− 6°C) or below in the field or garden. At the same time, it is important not to wait too long–the best advice is early spring planting, but after the weather has "settled." Late spring and early summer plantings generally are much riskier because dormant plants shipped in June are likely to arrive at your home in very poor condition. Do not plant this late.

Plant Arrival

Get plants as close to planting time as possible. When plants arrive, check the bundle. Each dormant plant should have a healthy crown (no leaves at this time of year) and a vigorous root system that is straw colored and reasonably fresh in appearance. Do not use old plants with black roots. A heavy covering of mold on strawberry roots and crowns indicates improper storage. If many of the plants in the bundle are moldy, discard the entire bundle. You can store plants which cannot be set immediately in a refrigerator for a few weeks, or until planting conditions are satisfactory. If necessary, moisten the roots, but do not soak them. Hold the plants as close to 34°F (1°C) as possible in the plastic bags in which they are shipped. Be sure the bags are closed by folding only; do not close them tightly.

Plant quality has a definite effect on plant establishment and yield. In the long run it is wiser and cheaper to purchase certified stock from a strawberry nursery than to secure planting stock from your own garden or a neighbor's planting.

Plant Distances

Strawberries are usually grown in the *matted-row* system in which runners are allowed to set in all directions (Figure 5). For this, set plants about 24 inches apart in rows 3 to 4 feet apart. Allow runners to develop and produce new plants to fill out the rows.

FIGURE 5. Strawberries are best planted in the matted row system. Spacing is 1 to 2 feet within the rows and 3 to 4 feet between the rows. Runners are allowed to set in all directions. Cultivation helps to straighten the runners into rows and to limit row width.

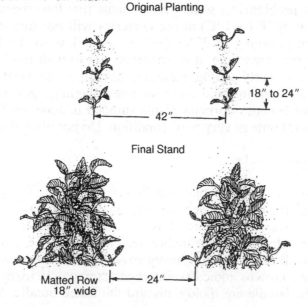

Original Planting

18" to 24"

42"

Final Stand

Matted Row
18" wide

24"

Generally, no effort is made to limit the number of runners produced, but they are kept within a row 18 to 24 inches wide. A matted-row 14 to 18 inches wide is preferable to one 24 inches wide, and the fruit is easier to pick and less subject to rot than the fruit that is in wide, crowded, matted-rows. Plants in the middle of traditional matted-rows widths (24 inches) are relatively unproductive.

There may be some advantage to adjusting in-row plant spacing distances on the basis of individual variety runnering habits. Varieties can be categorized as "prolific," "moderate," and "low," regarding runner production. Examples of "prolific" varieties would be 'Catskill,' 'Atlas,' 'Annapolis,' 'Cardinal,' 'Glooscap,' 'Veestar' and 'Sparkle.' Moderate types are 'Cavendish,' 'Earliglow,' 'Blomidon' and 'Jewel.' "Low" runner production varieties would be 'Kent,' 'Canoga' and 'Titan.' Relating this to within-row plant spacing, "prolific" varieties like 'Sparkle' can be stretched to 30 inches and 'Kent' can be condensed to around 15 inches.

Everbearer and dayneutral strawberry varieties are often grown in a *hill system* (Figure 6). This system is also useful for low runnering Junebearer varieties such as 'Titan.' Set plants about 10 to 18 inches apart in the row; at 10 inches remove all runners through the season. However, at 18 inches, it is advisable to hand set one runner between the mother plants (spaced row system). After planting, remove the first blossoms appearing on dayneutrals

FIGURE 6. In the hill system, plants are close together and runners are pruned off or prevented from rooting new plants during the summer. Plantings are usually made in double or triple rows in which plants are placed 12 to 18 inches apart, with a 20 to 24 inch alley between each double or triple set row. Fruit production is entirely dependent upon the yield of the single mother plant.

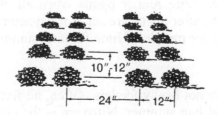

to encourage strong leaf and root development. Retain flowers appearing after the middle of June for later fruit in Northern U.S. and Canada.

Setting Plants

Transplant in the spring when the temperature is 40° to 50°F (4° to 10°C). Work the soil enough to obtain a moderately deep (6 to 8 inches) and reasonably loose planting bed. If the soil fertility was adjusted as recommended by a soil test, no additional fertilizer is needed until later in the season.

For ease in planting, cut back the ends of long roots, but do not root prune the entire plant. The fine roots will dry out in a few minutes on a sunny, windy day if the plants are not covered. If roots are dry, dip plants in water just before planting, but do not leave them sitting in water.

Set plants with roots straight down or slightly spread out but never bent. Cover the roots until the crown or stem, where the leaves arise, is just above the soil surface. If the crowns are covered with soil, or the roots exposed, plants will do poorly and may die (Figure 7).

Use a trowel, dibble, or other suitable tool to make holes for setting plants. Firm the soil about the roots with your foot.

After setting, give each plant a cup of water or turn on the sprinkler.

AFTER PLANTING

Blossom Removal

Blossom removal is a necessary facet of young plant care, whether the plants are Junebearers, everbearers, or dayneutrals. Plants grow better and produce more runner plants when all blossoms are removed a few weeks after plants are set. For dayneutrals, retain flowers appearing after the middle of June for later summer or fall fruit.

Fertilization

Strawberries are not heavy feeders. Often, no fertilizer is needed after planting until late summer. Nitrogen is the element most fre-

FIGURE 7. Correct planting depth for strawberry plants. Dormant plants with healthy roots should be used. The hole dug for each plant should be large enough to hold the roots without crowding. Plants with long roots should not lie even partially horizontal because this affects early and long term plant growth. After planting, tamp the soil firmly to remove air pockets around the roots. Water all new plantings well immediately after planting.

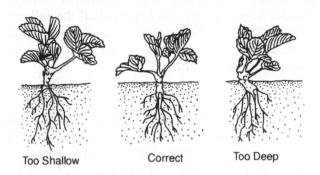

Too Shallow   Correct   Too Deep

quently needed. When the nitrogen supply is limited, the strawberry plants grow slowly and the leaves are small and light green. Extreme nitrogen starvation causes the leaves to become reddish brown. Excess nitrogen produces overvegetative plants with long petioles and dark, blue-green leaves. If the soil is sandy or if the plants are pale and lack vigor, apply one teaspoon of ammonium nitrate to each plant one month after planting. Place this at least four inches from the plant crown.

Potash is the second element most often needed in strawberry beds though it is rarely deficient in plantings following vegetable crops where large amounts of complete fertilizers were used. The deficiency produces plants with leaf margins which become reddish-purple to brown with a green triangle of color at the base of each leaflet. Potassium can be readily supplied through compost. If your preplant soil test indicates low potash, apply the material prior to planting and incorporate it into the soil.

Phosphorus is the third element most frequently added in supplemental fertilizers, but research indicates that the strawberry plant's requirement for phosphorus is very low, and no response to phosphorus treatment is likely.

Strawberries are likely to respond in late August or early September to an application of ammonium nitrate fertilizer at the rate of 3 pounds for each 100 linear feet of row. Spread it over the plants when the leaves are dry and brush it off. This fertilizer burns leaves if it is not brushed off quickly (with a broom), or if it is not washed off with water. No other feeding should be necessary except on very sandy soils, which may benefit from additional nitrogen in late winter. The rate suggested for this time is half the fall application. You will see nitrogen deficiency in strawberries developing if leaves are undersized and yellowish-green. In another month, the tips of the oldest leaves become red.

Cultivation

Shallow cultivation and hand hoeing are necessary to keep weeds from competing with strawberry plants for nutrients and water. Start early and repeat every ten days as long as weeds appear. If you are using a small tractor for cultivation, make the passes down the row in the same direction each time so that runner plants already rooted in a desirable position are not disturbed. Make successive passes farther from the original plants until the rows are the desired 14 to 18 inches in width. Some hand hoeing may also be necessary.

Many home fruit gardens are too large for hand weeding and too small for the use of heavy equipment. In many cases, herbicides can supplement these cultural practices to make the work of controlling weeds easier and faster. (Consult your county Cooperative Extension agent and read all labels closely.)

Watering

Adequate rains do not always occur when they are most needed during the spring and summer. If the weather has been dry, water several weeks after planting to stimulate new runner production and aid in their establishment. In addition, watering in late summer and fall can greatly promote flower bud formation. Irrigate when the amount of readily available moisture in the upper 6 to 9 inches of soil is 50% or less. You can determine this by the feel or appearance of the soil (Table 3). Apply enough water to penetrate the soil 6 to 8 inches. Sandy soils will require more frequent watering than heavier textured soils.

TABLE 3. Soil moisture determination.

Soil type	Amount of available moisture 50% or less	Field capacity
Sand	Appears dry; will not form a ball with pressure	Upon squeezing there is moisture but no free water
Sandy loam	Appears dry; will not form a ball with pressure	Same as sand
Clay loam	Somewhat crumbly but will hold together with pressure	Same as sand
Loam	Somewhat pliable; will form a ball under pressure	Same as sand

Adapted from R.W. Harris and R.H. Coppock. 1978. "Saving water in landscape irrigation," Leaflet 2976. Univ. of California, Div. Agr. Science.

Spacing Runner Plants

Runner plants can be spaced at definite distances by placing a small amount of soil just behind a runner plant to hold it in place until the desired number of runner plants is obtained for each mother plant. Usually 4 to 6 runners per linear foot of row is optimum with most matted row varieties. Crowded plants still yield well but produce small berries. Also, blossoms may not be pollinated well underneath the heavy plant canopy, and diseases and fruit rots will be more troublesome. Fairly narrow matted rows perform best (14 to 18 inches wide), but wide beds can be allowed if garden space is limited (no wider than 24 inches). After the desired plant density (4 to 6 plants per running foot of row) and row width are obtained in mid- to late-summer, remove extra runners every 10 days to 2 weeks for the balance of the growing season.

WINTER SEASON

The modern garden strawberry is raised throughout the world, including irrigated deserts in the Middle East, high elevations in the Tropics, and in extreme northern latitudes, including Alaska, where severe winters prevail. As grown in the colder winter climates of the northern United States and Canada, this herbaceous perennial depends upon snow cover and straw mulch to protect the roots and its crown. Left unprotected and exposed to winter's extremes, the crown is vulnerable to injury and death.

Mulching

Mulching not only protects the plants from winter injury, but it is also essential for clean fruit. A mulch can also suppress weed growth, conserve moisture and decrease losses from fruit rot. Cover plants with wheat, rye, or barley straw, or pine needles in late fall. The straw should be free of grain and weed seeds; if not, the bales can be left alongside the garden until the seeds sprout. The bales can also be moistened and covered with clear plastic to speed germination of the grain seeds. One bale of straw will cover about 100 square feet. Do not use hay because it contains too many grass and weed seeds to make a good mulch. Leaves, grass clippings, etc. are unsuitable because they will smother the strawberry plants. Oat straw will also pack and smother plants.

When to Apply Mulch

Apply the straw mulch after your area has experienced several light frosts but before temperatures drop to 20°F (-7°C). Temperatures below 20°F (-7°C) can cause injury. In Michigan or southern Ontario, straw is usually applied sometime after mid-November. In the mountains of North Carolina, night temperatures generally do not drop to 20°F (-7°C) until mid- to late-December. If the mulch is applied too soon, before plants are dormant, the straw can cause rotting of the leaves and crowns. If mulch is delayed too long, low temperatures could damage crowns.

How to Apply Mulch

Apply mulch 3 to 4 inches deep over the plant rows. It is easier to apply the mulch when the ground is frozen.

Mulch Removal

Remove the mulch in early spring as soon as there are signs of new leaf growth under it. Fork the mulch off the plants, and place it in the row aisles. Leave about ¼ of the mulch on the plants; the plants will grow through it. Straw placed between the rows will help smother weeds, conserve moisture, and help keep berries clean. Also, the winter mulch can be put back onto the plants during blossoming to protect strawberry flowers from frost/freeze conditions. The mulch traps heat from the soil and holds it around the plants. Mulch should not cover actively growing plants for more than a few days.

SPRING SEASON

Weeds

Some hoeing and hand weeding may be necessary in the spring. However, a good surface application of straw mulch should effectively control most weeds.

Fertilizer

Do not fertilize in the spring prior to harvest. Fertilizing at this time increases foliage growth, delays fruit ripening and increases berry softness and susceptibility to fruit rots (including gray mold and anthracnose).

Watering

Sufficient soil moisture is especially important during spring berry development for maximum berry size and yield and to extend the length of the harvest. Strawberries need about one inch of water a week, from bloom to the end of harvest. If natural rainfall is insufficient, make light applications of water during the heat of the day. Applying irrigations in the early to mid afternoon will allow the plant foliage and flowers to dry before evening. Gray mold infection is always a greater threat if plants are irrigated in the early evening. Excessive watering during harvest can cause soft, watery tasting berries.

Frost Control

Strawberry flowers, particularly of early varieties like 'Earliglow,' are quite susceptible to spring frost/freeze injury. A temperature of 30°F (-1°C) at plant level is likely to cause some injury to blossoms, but a temperature below 28°F (-2°C) will cause severe injury. Injury is more severe when low temperatures last for several hours. Unopened buds as they emerge from the crown are more hardy than open blossoms and, therefore, can tolerate slightly colder temperatures.

On small garden areas, a mulch cover can be an effective guard against frost/freeze damage. Place the straw or row cover over the plant during the daylight hours before a predicted frost, and remove the mulch or cover as soon as the low temperature threat has passed. Delaying the removal of the winter straw mulch slightly will serve to delay bloom and thus, possibly avoid frost/freeze injury in very early blooming varieties.

Considerable protection can be obtained by covering the plants with 2 to 3 inches of loose straw, or by covering with row covers made of spunbonded polyester or polypropylene (0.6 to 1.0 oz. per square yard). Keeping plants wet through the night will also give protection since the change of water to ice on the plant releases heat. The continuous addition of water allows the continuous change into ice and, therefore, the continuous generation of heat. A milky white ice coating indicates that not enough water is being applied; clear ice indicates adequate sprinkling. Use sprinklers that deliver about 1/10th inch of water per hour. You can catch the irrigation water in a rain gauge to see if the sprinklers are applying about 1/10th inch per hour. Begin sprinkling when some ice is detected on the strawberry leaves, and continue irrigating as long as ice forms. Sprinkling must continue until the ice coating begins to melt in the morning and continues to melt when no new water is applied. Do not attempt this method if wind speeds (at the ground level) are above 5 miles per hour.

Pollination

Most strawberry varieties (except 'Apollo') are self-fruitful, but cross-pollination between strawberry varieties usually results in better fruit set, larger berries, and fewer misshapen or "catfaced" berries. Pollination automatically occurs by wind-motion in strawberries, but honeybees are the best pollinators. Bees can help to increase berry size by 20%, and reduce the amount of malformed fruit by more than 25%. Other insects that are effective in pollinating strawberries include blowflies and syrphids. Never use any insecticides during the bloom period.

Harvesting

The first harvest can be made about 30 days after the first bloom. Not all berries ripen at the same time, and picking may continue for two to three weeks for each variety grown. Pick only those berries that are completely red, and leave berries showing any white for later harvests. Harvest the planting every two to three days. During hot weather, you may have to pick everyday.

Pick the berries with the cap or stems attached. Remember to break the stem by pinching it between the thumb and forefinger. This will prevent damage to the plant and keep your berry from bruising. Be sure to "pick out" berries that are rotting or have been damaged by slugs, insects or birds (they all love strawberries, too). Strawberries are highly perishable and should be used soon after harvest. Keep harvested berries out of the sun and refrigerate as soon as possible after picking. Do not wash strawberries until you are ready to use them. Instead, clean them by gently dropping the berries into a basin of cold water. Hull berries *after* you have cleaned them, using a small blunt knife. Freezing whole strawberries is not satisfactory. Cut them in half, and slice or puree them before freezing.

Yield

The actual yield depends upon many factors, such as the variety, the fertility of the soil, the training system used, and whether "certified" disease-free nursery plants are used. You can expect nearly a quart of fruit per linear foot of row if you follow good cultural frost and pest management practices. Well established, weed-free matted row plantings can produce berries for up to three years. Everbearing varieties and dayneutrals do not produce as large a spring or early summer crop as Junebearers, but the combined spring/early summer and fall crop should give good yields.

RENEWING A PLANTING: MATTED ROW STRAWBERRY

As soon as picking is completed, apply a 10-10-10 fertilizer at the rate of four pounds for each hundred feet of row. Scatter the material over the tops of the plants while they are dry and use a broom to brush it off the foliage.

Set the blade height on your rotary mower to 4 inches and mow the leaves from your plants. Be careful to rake and dispose of them; they may carry diseases and insects. Do not mow in northern areas of Canada or Alaska, where berries ripen late and the growing season is short.

Narrow rows with a cultivator, rototiller, or hoe to a strip 12 to 18 inches wide. Then, thin the plants to about six inches apart in all directions, leaving only the most healthy and vigorous.

After mowing, narrowing and thinning beds, water the plants to carry the fertilizer down to the roots and to encourage new plant growth. Fertilizing, mowing and watering are done to encourage plants to resume active growth after the harvest.

Apply ammonium nitrate in late August or September as outlined previously. Also, apply straw mulch as recommended under "Winter season."

VARIATIONS IN MATTED ROW CULTURE

Matted row is the simplest form of strawberry culture and the most economical for the northern United States and Canada.

One important modification for milder areas of the South (not subject to winter minimum temperatures below 10°F ($-12°C$)) is that winter mulch protection may not be necessary or advisable. In North Carolina, for example, winter straw mulch application is recommended in the mountains of western North Carolina where the climate is similar to that of the Middle Atlantic States. However, in the milder climate of the central and eastern sections of that state, strawberry plants are not covered in winter. In these areas, a coarse application of wheat straw or pine straw is made before growth starts in the spring primarily for keeping the fruit clean.

Matted row varieties will also vary considerably from north to south (and east to west in Virginia and North Carolina). However, most of the recommendations for matted row or matted-bed culture do not vary much between northern and southern areas. In milder climatic areas such as eastern parts of Virginia and North Carolina, matted row beds can be transplanted from February to early March while in western Virginia and western North Carolina, early April transplanting has given the best results.

ALTERNATIVE CULTURAL SYSTEMS

Fall-Planted Annual Hill

This system is for the warmer areas of the South which have mild winter seasons where temperatures seldom go below $10°F$ ($-12°C$). The success of the fall-planted annual hill system depends heavily upon how much growth the plants can make during the winter months. If the winter is warm, production can be as high as one pound per plant; if the winter is cool, or if the planting is not protected from late winter/early spring frosts, yields can be as low as 1/4 pound per plant or about 25 pounds for 100 plants. Two types of plants can be set in fall-planted annual hill beds:

1. Fresh, fall-dug nursery plants.
2. Another plant that has recently been introduced to commercial growers that is called a "plug plant."

The primary advantage of the plug plant is that it does not require constant overhead irrigation during the first week after setting as does the highly perishable bare root fresh dug plant. Plug plants are rooted from strawberry plantlets or tips in cylindrical trays with a daily misting cycle. After misting for 7 to 12 days, the plugs are moved to a fully exposed nursery gravel pad for another 2 to 2 1/2 weeks of growth and acclimation before transplanting. Plugs are generally more expensive to purchase than fresh dug plants. Transplanting dates for plugs can be slightly later than for fresh dugs without as great a yield reduction. This is because plugs establish more quickly than fresh dugs after transplanting.

In the fall-planted annual hill system, plants are typically set from late September (e.g., eastern Virginia) through early November (e.g., Florida and South Texas), depending upon region. Prior to planting, nine inch high raised beds are formed. Usually a two-row bed with a spacing of 12 to 15 inches between rows and between plants down the row is used.

Plants are hand-planted through the black plastic mulch in 1 1/2 inch slits in the plastic. These are cut by trowels or special spacing wheels. Use only undamaged, non-dormant fresh dugs or plug plants with good leaves. For fresh dugs, begin watering immediately after transplanting. These plants will require intermittent watering for several days. Each day, start watering in the morning when plants show moderate wilt and continue until the hot part of the day has passed. Only a small volume of water is required (1/10 inch per hour). Plug plants require watering at transplanting and during any dry spells in the fall following transplanting.

The major advantages of the high-density fall-planted annual hill system in the South are uniform plant stands unaffected by summer diseases, drought or weed competition; earlier fruit harvest (10 days to 2 weeks); easier harvest; and shorter turnaround time from planting to harvest (6 months). The two main varieties recommended for home garden "annual hill plasticulture" are 'Chandler' and 'Sweet Charlie.' More details about these varieties and this system of culture can be obtained from your state Cooperative Extension service, or by contacting E.B. Poling, Box 7609, North Carolina State University, Raleigh, NC 27695-7609 (ask for Horticulture Information Leaflet 205-G, Annual Hill Plasticulture for the Home Garden).

PROPAGATION

Dormant, cold-stored plants are recommended for matted row planting systems that require spring or early-summer transplanting. Dormant plants are not recommended for fall-planted annual hill plasticulture. Leafy, non-dormant, "fresh-dug" plants may be obtained or dug by the home gardener during late summer for fall transplanting in hills. Also, plug plants are highly recommended for annual hill plasticulture because of their ease of transplanting. However, strawberry propagation is best not attempted by the home gardener. Infection of home garden strawberries by virus diseases over time is one of the important reasons to buy certified virus-free stock for all new plantings.

EVERBEARING BERRIES

Everbearing strawberries are grown primarily for the fall crop. They will produce satisfactorily if grown under the spaced-plant system of culture. Successful production of this type strawberry requires much labor, so planting should be of limited size. They will not do well when grown in matted rows.

Prepare the site and set the plants as you would Junebearing varieties.

Maintain the plantings under either a sawdust or a black plastic mulch. With a sawdust mulch, care for the planting as if it were a regular planting until early June when runners appear, then stop cultivation. Fertilize each plant with 1 tablespoon of a 10% nitrogen fertilizer or equivalent, spreading the fertilizer uniformly over the soil around each plant.

Cover the entire planting with one inch of either hardwood or softwood sawdust. It may be fresh or weathered. Don't apply excessive amounts. Further weed control must be done by hand, since hoeing and cultivation will mix the sawdust with the soil, thus destroying the mulch benefits.

After applying the mulch, start training the runner plants, locating each in the desired position. Force the plants gently but firmly through the sawdust so their roots contact the soil. The distance between runner plants varies from season to season but will be about 8 to 10 inches.

After the desired number of runner plants has been established (4 to 6 per linear foot of row), remove all others as they develop through the remainder of the season.

Continue removing flowers until the first to the middle of July. The exact date for discontinuing blossom removal depends upon the planting. The more vigorous the plant, the earlier blossom removal can be stopped.

Begin harvesting about 30 days after the first blossoms appear. The first berries will ripen in August and harvesting should continue twice a week until frost.

Black plastic offers advantages over sawdust as a mulch for everbearing strawberries. It minimizes weed problems and helps keep the berries cleaner. With black plastic, only a slight variation in cultural practice is needed. The mulch may be spread over the row area, and the plants set through it at desired locations. Cover edges of the mulch with soil.

Establish runner plants where needed by cutting a slit in the plastic and placing the plant firmly into the soil. Blossom and runner removal is the same under both mulch systems.

Trickle type irrigation lines or soaker hoses installed under the plastic mulch can prove helpful during drought periods. Take care not to over water.

Junebearing strawberries may be grown according to the spaced plant systems, too, but benefits do not justify the added efforts.

CONTAINERS

Strawberries may also be grown in pots and planters to add a decorative touch to decks and patios. Special terra-cotta or earthenware planters, with little pockets in their sides, allow the plant to trail and root. To set strawberries in these containers, first line the base with rocks or stones to facilitate drainage. Then gradually fill the container with a good quality potting mix.

PESTS

Above all, it is important to protect your strawberry plants and fruits from insects and diseases. Protection should also be provided from snails, slugs, mice, and birds.

Insect and Disease Problems

Although strawberries can have their share of insect and disease problems, most homeowners ignore them unless they become serious. Following these six precautions should minimize pest problems.

1. Use only certified, virus-free plants for setting.
2. Choose well-drained soil; follow rotation recommendations and have a nematode assay made.
3. During harvest remove berries damaged by diseases and insects as this reduces the amount of fruit rot.
4. Properly renovate beds to remove older diseased foliage and keep from getting too crowded.
5. Don't keep a planting in production too long; start a new planting every year or two to replace old plantings after their second or third crop.
6. Do no allow insects and diseases to build up.

Should a serious problem develop, your county Cooperative Extension agent can give information on the latest, safest, and most effective chemicals to use. Ready-mixed commercial packages of pesticides can be purchased under various brand names, or the separate materials can be purchased and combined. In either case, read the labels on containers to determine contents and directions for use.

Protection from Birds

Birds like strawberries too, especially since they are ripe when few other fresh foods are available to them. Protect your berry patch by covering it with bird netting. It's easy to remove when harvesting and can be used year after year. Be sure to anchor edges so birds don't walk under netting. A permanent removable cover (for beds and strawberry pyramids) can be built from chicken wire tacked to a board frame.

Scaring devices (such as aluminum whirling devices) sometimes keep birds away. If you use these devices, put them into service early in the season at the first sign of fruit ripening and before the birds have become established in a feeding area. These devices must be operated from dawn to dark and moved around frequently before the birds become accustomed to their noise emitted from a particular location.

Index

Page numbers in *italics* indicate figures; page numbers followed by t indicate tables.

T - #0507 - 101024 - C0 - 229/152/15 - PB - 9781560220572 - Gloss Lamination